T0269079

This book explores the many engineering and architectural aspects of submarine design and how they relate to each other and the operational performance required of the vessel. Concepts of hydrodynamics, structure, powering and dynamics are explained, in addition to architectural considerations which bear on the submarine design process. The interplay between these aspects of design is given particular attention, and a final chapter is devoted to the generation of the concept design for the submarine as a whole. Submarine design makes extensive use of computer aids, and examples of algorithms used in concept design are given. The emphasis in the book is on providing engineering insight as well as an understanding of the intricacies of the submarine design process. It will serve as a text for students and as a reference manual for practising engineers and designers.

CAMBRIDGE OCEAN TECHNOLOGY SERIES

General Editors: I. Dyer, R. Eatock Taylor, J. N. Newman and
W. G. Price

CONCEPTS IN SUBMARINE DESIGN

CONCEPTS IN SUBMARINE DESIGN

Roy Burcher
and
Louis Rydill
Department of Mechanical Engineering, University College London

CAMBRIDGE
UNIVERSITY PRESS

CAMBRIDGE UNIVERSITY PRESS
Cambridge, New York, Melbourne, Madrid, Cape Town,
Singapore, São Paulo, Delhi, Tokyo, Mexico City

Cambridge University Press
The Edinburgh Building, Cambridge CB2 8RU, UK

Published in the United States of America by Cambridge University Press, New York

www.cambridge.org
Information on this title: www.cambridge.org/9780521559263

First published 1994
First paperback edition 1995
Reprinted 1998

A catalogue record for this publication is available from the British Library

Library of Congress Cataloguing in Publication data

Burcher, Roy.
Concepts in submarine design / Roy Burcher and Louis Rydill.
  p. cm. – (Cambridge ocean technology series; 2)
Includes bibliographical references and index.
ISBN 0-521-41681-7 hardback
1. Submarine boats – Design and construction. 1. Rydill, Louis.
II. Title. III. Series.
VM365.B89 1994
623.8'205–dc20   93-2614 CIP

ISBN 978-0-521-41681-8 Hardback
ISBN 978-0-521-55926-3 Paperback

CONTENTS

INTRODUCTION

The two authors of this book have been involved in the design of submarines for the Royal Navy for upwards of thirty years, and have also been involved on and off in the teaching of submarine design for much of that time. They both have connections with Vickers Shipbuilding and Engineering Limited (VSEL) the only builders of submarines in the U.K., Louis Rydill as a design consultant and Roy Burcher as the VSEL Professor of Subsea Design and Engineering at University College London (UCL). Roy Burcher runs a postgraduate design course at UCL, which is attended by students from many countries.

With this background, we are only too well aware of the dearth of textbooks on submarine design and engineering. There are also relatively few technical papers on the subject. There was a seminal paper in 1960, published by the Society of Naval Architects and Marine Engineers of the U.S.A., entitled *Naval Architectural Aspects of Submarine Design* by Arentzen and Mandel, which we regard as an outstanding contribution to the subject but, no doubt because of security issues involved in military submarine design and operation, that splendid opening up of the vistas made possible by the advent of nuclear propulsion for submarines has subsequently become largely closed off to view.

Yet there is still much about submarine design and engineering which can be said without risk of offending against security obligations. The course at UCL is, in fact, completely unclassified and is open to all prospective students with the appropriate qualifications. This book draws on the authors' experience in devising and mounting that course, without going into either the detailed theoretical source material provided or the computer programmes available, which are invaluable in enabling students' submarine exercises to proceed so far in concept design evaluation in a matter of several weeks.

The book is primarily intended to serve as an introduction to the process of designing a submarine by providing a grounding in the principles involved. The emphasis is more on creation and synthesis of concepts and less on methods of analysis, though some aspects of the analysis are treated where we consider they are necessary for an understanding of the factors which shape a submarine design. The book should be suitable for naval architects and engineers with a marine background who are about to embark on a career which might take them into the fields of submarine design and engineering, construction, operation and support in service. We would wish to emphasise that it is at least as important for those engineers who might be involved in any of those activities to appreciate the rather special considerations which determine why submarines are configured the way they are and operate in the way they do, as it is for naval architects working more directly on design.

Because our experience is with military submarines we have focused on them in our text. We consider that much of what we say can be applied to any manned underwater vehicle, but suggest that the design and engineering considerations relevant to unmanned underwater vehicles are sufficiently different to necessitate caution in extrapolation from the technical base we describe.

It will be observed that we have chosen to write throughout the book in the first person/plural mode. This is not only in the hope of avoiding the stiffness that can result with the impersonal mode, but also to convey that much of what we express is our opinion of the way things are (or ought to be) in submarine design.

FORMAT OF THE BOOK

We encountered some difficulty in structuring the contents of the book. This was because each aspect is so closely interrelated to the others that it is difficult to treat any one in isolation. For that reason there is some element of repetition as the same topic arises in different contexts.

We start with a chapter on the general principles and structure of the design process which apply to any large engineering project but with submarine design as the principal objective. Succeeding chapters approach the design problem from different points of view so that, hopefully, a grasp of the variety of considerations is obtained before returning to the approach to Concept Design of a submarine.

In the second chapter we take a brief look at the history of the submarine from a particular perspective: those vessels which we believe represent milestones marking out the evolution of the submarine towards its present day manifestations.

We then turn to the basis of naval architecture – the consideration of the hydrostatics of a submarine which has to operate both at the surface and fully submerged. This poses constraints which permeate the whole of submarine design.

The next chapter is allied to the previous one in that it looks at the disposition of weight and space and how these considerations govern the sizing of the submarine.

As a unique feature of the submarine we go on to discuss the diving capability of the vessel and the configuration aspects of pressure hull structure to withstand sea pressure at depth. We have tried to convey an understanding of the function of the structure whilst deliberately excluding much detail of structural analysis which is a specialist topic in its own right. For those who wish to go further there are a number of references listed at the end of the book.

The next subject considered is the powering of the submarine. In our context this includes a discussion of speed requirements and submerged endurance. The search for energy economy leads to preferred shapes of the outer hull form and the selection of propulsion devices.

The demands of structure and powering in the configuration of the hull lead in the next chapter to a discussion of the architecture of a submarine – how best to arrange the various demands for space within the close confines of the hull geometry.

A submarine is free to move in all directions when submerged but there are limits, and the next chapter gives consideration to the control of a submarine over the range of speeds. This is dictated by the geometry adopted and leads to possible limits of operational freedom in the interests of safety.

The operation of a submarine requires special systems to be fitted. A chapter is devoted to the considerations that lead to the sizing and disposition of these systems so that due allowance can be made for them in the design. The overall designer must allocate adequate budgets and define the duties so that the specialist designers can provide the required systems.

The aspects of how the submarine is to be built and where costs arise are then addressed. Though only indirectly influencing the submarine design process, they are important aspects to be borne in mind throughout the design.

In the final chapter we return to the process of generating concept designs of a submarine to meet the operational requirements. A number of approaches are suggested by which a designer, though addressing them individually, might use more than one in arriving at a tenable solution. We argue against any rigidity or rule formulation for this process as this can inhibit the freedom to innovate which is an essential feature of Concept Design.

In the course of writing the chapters we have generated some material which, though relevant, represented diversions from our main theme. In some cases these are explanations or descriptions whose value would depend on the background of the reader. These have been included as Appendices.

Finally we would observe that throughout the book we make reference to 'the designer'. This term is used for convenience but should not be taken literally. The design of a submarine involves the efforts of many engineers and scientists and so we use the term to represent the team which may vary in number during the process of design and building.

ACKNOWLEDGEMENTS

Whilst the views and ideas expressed in this book are the responsibility of the authors, we are indebted to the many people who have contributed to the Submarine Design Course at University College London over a period of some 20 years. However, in identifying those contributions we have necessarily had to work on the basis of more recent involvement with the course.

The course requires a considerable effort in day to day running and in the provision of computer programs and data base. John van Griethuysen and Simon Rusling provided a considerable input to these activities.

A number of external lecturers are invited to present their specialist knowledge and some of their information is incorporated in various ways throughout the book. In that way, contributions have been made by: Philip Dent and John Charlesworth of DRA Haslar on Hydrodynamics; Matt Roberts and David Cresswell of DRA Dunfermline on Structures; Andy Stevens on AIP Systems; and Vernon Dawe on Electrical Systems (which is not the forte of the authors); Rod Puddock and Paul Wrobel on the Building and Construction from the shipbuilders standpoint. We appreciate the help of Richard Compton-Hall, Director of the Submarine Museum, for providing some of the information on earlier submarines.

That the course ever existed is due to the support given by the Heads of the Royal Corps of Naval Constructors over many years. We wish to express our gratitude to Noel Davies and Tony Peak, Chief Executive and Deputy Chief Executive of VSEL, for their sponsorship of the VSEL Chair at UCL, a significant contribution to the furtherance of the teaching of Submarine Design.

Finally, a deep debt of gratitude is due to Gill Scoot without whose endless patience and skill, through many drafts, the book would not have been created at all.

1 DESIGN IN GENERAL

INTRODUCTION

In this book we attempt to convey an understanding of those aspects of design which are specific to submarines. However, submarine design is just one of many engineering design activities and there are general features which are common to all. Before proceeding to the specific aspects of submarine design, we consider it worthwhile in this first chapter to address the more general aspects and show how they relate to the submarine design task.

DESIGN OBJECTIVES

1.1 Though there are many variations on what may be considered design objectives we suggest that the following three are primary in all designs and should be sustained throughout the whole design process:

    (a) that the product should perform the functional purpose of the customer or operator;

    (b) that the design should be suitable for construction within the capability of the technology and resources available;

    (c) that the cost should be acceptable to the customer.

Though expressed as separate objectives they are interactive and may on occasion be incompatible. The circumstances within which the design takes place may lead to one or another becoming the prime objective with the others subsidiary.

In some situations the performance of the design is paramount: only a design that is capable of fully performing the required function is of interest. In such circumstances the customer would have to be prepared to pay the cost of such a design and even the capital investment to create the technology and resources necessary to realise the design. Some major nations have accepted this situation for defence equipment in the past but it is less prevalent in current political circumstances.

The technology and resource objective may become dominant, particularly where a design is being generated for production in a country other than that of the designer. He must therefore recognise limitations in resources which do not exist in his own country or company.

The produceability of the design is important both to cost and time to

produce. The designer must in consequence be well-informed on manufacturing issues and understand where the limits lie. Failure to take
account of production processes can be expensive: for example, an early
version of box girder bridge, though structurally adequate in its final
form, failed due to the different loading conditions during construction;
also the original design of Sydney Harbour Opera House posed difficulties in construction that necessitated considerable changes from the original concept. In ship and submarine design, the aspect of buildability can
take the form of avoiding overly compact arrangements (sometimes
advocated in the interests of keeping down size in the expectation that
this will keep down cost) if, in the outcome, difficulty of access to construct and fit out the vessel causes cost to increase.

In other circumstances the cost becomes dominant, because there is
some form of budget limit. In virtually all designs some form of value for
money or cost effectiveness will operate as a criterion for acceptability.
There may be a ceiling on cost which will eliminate even cost effective
designs if they exceed this limitation. Some projects have been run on a
'Design to Cost' basis.

For the majority of environments within which design is conducted all
three objectives are considered important without any one being dominant. It then behoves the designer to attempt to create a design which
has satisfactory performance, can be built and is acceptable in cost.

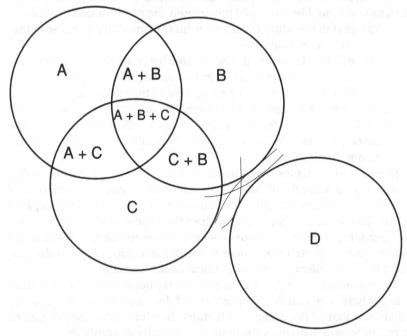

Fig. 1.1   Fuzzy set analogy

If an acceptable solution is to be found it is imperative that a close and continuous dialogue should exist between the designer and user or operator during the initial phases of design. The designer must fully understand the aspirations of the operator and equally the operator must understand the implications and consequences of his requirements.

1.2 This dialogue usually leads to what is frequently referred to as 'compromise' in design. The authors dislike this term as it implies an end result which is 'less good' than it might otherwise have been. We would prefer to consider that the design is an optimum solution in the circumstances which prevail. To illustrate this one can use the methods of set theory and diagrams. Say we plot three non coincident points A, B and C as representing three requirements of the design and their location in parametric space. About each point we draw circles to represent a degradation in individual performance, say of 70% (Figure 1.1). These circles may overlap each other, in which case the designer has some options on where to aim his design. He may choose the zone (A + B + C) so that each requirement is achieved at slightly better than 70%. Alternatively he may aim for the zones A + B, B + C or C + A, in which case somewhat better performance is achieved in two out of three at the expense of poorer performance in the other feature. He may also aim for very high performance in one requirement A or B or C at the expense of performance in the other two. The requirement D poses a real difficulty as it has no overlap even at 70%.

1.3 Another notion in design is associated with these three broad objectives and that is 'interaction', for example, the search for high capa-

Fig. 1.2   Jigsaw analogy

bility with resources and costs, usually raising their threshold. Also resource limitation will interact with performance and so will cost ceilings. As should become apparent in later chapters, 'interactions' occur in all aspects of design. For submarines they persist to very detailed levels of design so that almost every decision has far reaching implications for other features of the design; occasionally, unanticipated repercussions might act to render the whole design invalid.

A good design may be considered to be like a jig-saw with all the pieces (component systems) closely interlocking to form a whole picture (Figure 1.2). If it is now imagined that one piece changes shape it will no longer fit, and so its immediately adjacent pieces have to be changed to allow it to fit. These consequent changes bring about dislocations with other pieces and this process may continue until nearly all the pieces have to be modified to achieve a new picture. More tolerant designs, such as most surface ships, would involve changes to far fewer pieces.

DESIGN PROGRESSION

1.4   The design process is usually considered to start with the definition of requirements and end with a complete set of production drawings and specifications for construction. It may, however, be argued that even this is not the end of design and that it continues through building of the first vessel and subsequent trials. It is a fact that these latter stages do often result in modification to the design, the information on which is fed back to later vessels of the class and to provide data for future designs. However, for purposes of discussion we propose to consider the production information as the end point.

The achievement of an end product requires a structured progression of design with definite stage and decision points on whether to proceed. This introduces the idea that the design process is one of increasing

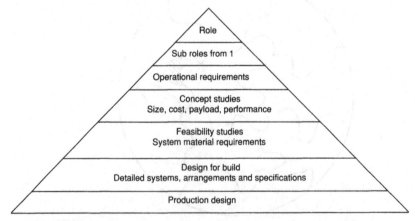

Fig. 1.3   Design pyramid

detailed definition. It may be likened to a layered pyramid (Figure 1.3). At the apex there is a very broad, simple description and at the base is the fully detailed production definition. Each layer contains a definition of the whole vessel but of differing detail and character.

Thus the apex may be considered to be a definition of the role of the submarine. We assume that the designer's task starts when it has been decided that a submarine solution is required, i.e. that there will have been higher level studies of alternative methods of meeting an operational requirement of which the submarine solution is only one. This first level description of the submarine therefore takes the form of a description of its task. For example, it may be required to conduct a barrier surveillance operation in the North Atlantic or, for an entirely different type of vessel, to conduct search and salvage operations on the sea bed in some defined ocean area. The next level of definition is an expansion of the role description, though still in operational terms. This would include the area of surveillance to be covered and the associated ocean climatic conditions, time expected to be on station and the expected levels of reliability to perform its role. This level of definition represents a first stage in that it should be entirely generated by the customer or, more usually, by his operational staff.

It is at the next stage that dialogue between operators and designers commences. The definition changes from operational description to a more specific performance description. This description may be subdivided into two parts. To perform its task the submarine will require equipments, sensors, communications, weapons or other prosecution methods such as manipulators for salvage. The performance characteristics of these equipments need to be specified. At the same level, the performance characteristics of the submarine as a vehicle need to be specified, how far it has to transit, how long should it take, how fast, agile or what precision of position is required, how deep must it dive and how long should it remain dived.

1.5   From this performance description the designer can begin to generate Concept Studies. These will be broad brush technical descriptions of possible solutions to the requirements and will constitute the next level of description of the submarine. The description will include a selection of equipments and associated manning (the Payload) and estimates of size, power, configuration and cost of the vehicle to carry them.

As would be expected many such concept designs may be generated. They could be technically straightforward designs of varying capability and cost which arise because of the interaction effects previously discussed. Incompatibilities may occur between equipment and vehicle performance and lead to various trade-off studies to seek an optimal solution.

The generation of such Concept Studies is primarily an act of synthesis. In the final chapter of this book, we discuss some formalised approaches to generating concept designs to create the basic parameters from, as it were, a blank sheet of paper. However, too formal a method can stifle innovation and new ideas which are the essence of concept generation. At the concept stage there is scope for imagination and innovation, though the resultant designs should have some probability that they can be physically realised. For that to occur the designer must have a thorough knowledge of all aspects of submarine systems and how they interrelate. The subsequent chapters of this book treat the major subsystems of a submarine with the object of conveying that understanding before we return to the concept design process.

The Concept Studies form a basis for a dialogue with the operators and the possibility of trade offs. It is probable that there will be various degrees of risk as well as ranges of performance and cost involved in the concepts. Those with a higher degree of risk should be utilised to provide a basis for research and development programmes if they offer substantial benefit over more straightforward designs.

1.6   At the end of the Concept Study phase a few of the more promising designs will be selected for further study. The design process now moves to the next phase of Feasibility Study. Though there may still be some scope for innovation, this stage is characterised primarily by analysis rather than synthesis. From the basis of the outline description of a selected Concept Design it is possible to conduct more detailed engineering analyses to confirm the basic features of the design. Layout drawings will be prepared to establish that all the equipment, accommodation and systems can be fitted into the proposed configuration of the hull. Detailed structural analysis of the pressure hull will be performed and a more extensive analysis of power and performance will be conducted. These analyses provide a basis for more accurate assessments of the weights and a comparison with the buoyancy elements will indicate whether the hydrostatic balance can be achieved with adequate margins. Considerably more effort is required and the single designer (or small team) associated with Concept Studies needs to be expanded and should include members with more specialist knowledge of the primary systems.

The Feasibility Studies will generate a considerably more detailed definition of the design which may be regarded as constituting another level of the design pyramid. An important output of this stage will be the generation of 'budgets' in terms of weight, space and power for the various sub-systems of the design. The budgets are important in that they provide a basis for the control of the design should it be decided to proceed further with the design option.

It is highly probable that the outcome of the Feasibility Study phase will be to reveal some deficiencies in the original concept. If the Concept Designs have been created by knowledgeable designers the shortfalls revealed by Feasibility Design should be relatively modest. Nevertheless, there is likely to be some need to revise the original configuration, weight, performance and cost assessments. Thus some reiteration of the embryonic design is usually necessary before it can be considered viable. This iteration process is a common feature of all engineering design. It also brings with it another common feature which might be termed a self learning or heuristic process. The act of exploring the various features of a design leads to a greater understanding of the interaction processes at work within the design. Though broad guidelines are offered in this book it is our experience that every design has its own particular features whereby certain aspects are sensitive to change and others seem to make little difference. In the broadest sense, some designs are found to be volume critical and therefore highly sensitive to spatial arrangements, whereas others are weight critical so that small increases in system weight have an inordinate impact on size.

Another characteristic of submarine design is the multiple functionality of many systems. One function may so dominate the design of a particular system that changes in other functional requirements have little effect. As an example, consider an internal main transverse bulkhead: it performs the function of supporting the pressure hull shell; it may also provide an escape subdivision within the hull and therefore have to withstand lateral pressure; it may also provide support of the weight of major items of equipment within the hull. Any one of these may dominate the bulkhead structural design in such a way that it can easily perform the other functions and not require any revision if these secondary functions were subject to small changes.

1.7   On completion of the Feasibility Studies a decision will be made on whether to proceed with the design. The most promising of the designs studied will be selected for carrying forward to the next stage of design.

No common terminology is in use for this further stage but it may appropriately enough be described as Design for Build. The objective of the stage is to develop the selected Feasibility Study design in sufficient detail to be offered for Tender to Build and for the production drawings to be produced.

The depth to which the design will be taken will vary depending on the way in which the contract for the construction of the first submarine is to be placed. If it is both for design and build then much of the work will be carried out by the successful tenderer for the contract, e.g. the selected builder. If the contract is only for build then the design work for this stage will be conducted by an 'in house' design team prior to issue of an

Invitation to Tender. Whichever way the contract is placed there will necessarily have to be a more detailed description of the design at the lower level of the pyramid in Figure 1.3, which we have termed Design for Build.

The development of this greater detail entails a considerable expansion in the size of the design team with the inclusion of many specialist designers in such areas as weapon systems, electrical systems, other oil air or water systems, structural designers and machinery specialists. Whereas the small team involved in Concept and Feasibility Studies can usually retain an overview of the totality of the design and be aware of the consequences of changes, this perspective is difficult to retain with the many specialist groups.

To retain control of the design and ensure that it conforms to the original intent much greater discipline and coordination needs to be imposed. The best way to instil this discipline is for the Design Manager to allocate budgets to the separate system groups. The budgets should not be simply cost allocations but take the form of multi-parameter bounds within which a specialist group can work. To ensure retention of the earlier hydrostatic assumptions, budgets of weight, spatial location volumes and centres of gravity are required. To ensure the overall power and energy assumptions are sustained, allocations of power and centres of distribution need to be made.

These budgetary allocations are an important outcome of the Feasibility Studies. By regular monitoring of the budgets the Design Manager can keep track of the state of the design and be alerted to possible situations that may potentially endanger the successful outcome of the design. If a specialist group encounters difficulties in remaining within budget, the design manager must try to resolve this situation. Providing the original estimates were realistic and adequate design margins allowed, resolution may be achieved by a revision of budgets on the basis that some have proved to be more generous than others so that a trade off can be made. In other, more serious, situations some revision of the estimated performance may be called for, which amounts to a degradation of the approved design. If this could not be accepted a more major revision of the design would probably be necessary.

In a sense, a set-back of that magnitude would amount to a failure of the original Concept and Feasibility Designs and, awareness of the risk, reinforces the need for the designer at the earlier stages to be fully acquainted with the subsequent design interactions that will arise, even though he would not be able to pursue them in detail. Thus, for example, he must appreciate the design implications of providing a hydraulic power plant and distribution system even though they will not be designed until much later.

The greater the degree of innovation in the Concept the more likely are design impasses to be identified in the later stages of design because less is known about the consequential design interactions than with a more conservative design based on past building. This should not be taken as advice to avoid innovation but rather to alert the prospective designer to its dangers. Whilst excessive margins are undesirable an innovative Concept merits having greater margins applied if it is to have the best chance of succeeding. It is only when the design is developed in detail that problems will be revealed, which is an aspect of the self learning process mentioned earlier. If genuinely innovative advances are to be made without undue risk there is a case in favour of design teams taking such ideas through to the Design for Build stage, even if there were no current intention to place an order for build. In that way a design team would learn more about the problems and consequences and be better prepared for the future when an intent to build does exist.

1.8 By the completion of Design for Build the following levels of detail may be expected to have been achieved.

Structure
Main Pressure Hull configuration with thickness of plating, spacing and geometry of frames and dome closures.
Main internal structure scantlings for tankage, bulkheads and main decks.
External tankage and free flood structure including Bridge Fin.
Details of major structural connections and hatch penetrations.
Major seatings for large items of equipment.
Specification of welding procedures and non-destructive testing of fabrications.

Arrangements
General Arrangement drawings developed for all compartments and deck layouts.
All major items of equipment and major systems located on plans by compartment layouts.
Layouts of bunk spaces and arrangement of cabinets and consoles.
Detailed arrangements of sensors and masts.
Detailed arrangements of weapon launch system and reload stowage.

Hydrodynamics
Resistance and Powering estimates confirmed or revised by model tests.
Control and dynamic stability confirmed by tests and main control surfaces sized.
Power demands defined.
Design of propulsor developed in detail including model tests.

Systems

Centralised power plants sized for capacity and suitable equipments selected, (pumps, motors, compressors, bottles).

Schematics of system runs developed and main runs of piping and cables located in compartments.

Pipes and cables sized and main valves, switches and bulkhead penetrations defined and selected.

Appropriate standards and specifications selected.

Hydrostatics

All contributions to buoyancy calculated with determination of vertical and longitudinal centres.

Majority of weights calculated in detail and centres of gravity calculated.

Main Ballast Tanks and Trim and Compensating Tanks sized and located.

Permanent Ballast calculated and located to give required longitudinal balance and transverse stability, both surfaced and submerged.

1.9   As will be seen from Figure 1.3, there is a further activity level in the Design Pyramid and that is Production Design. In all cases this activity, in which every component of the vessel must be defined for manufacture, installation and testing, will be undertaken by the submarine builder. It will be appreciated that Production Design is a considerable undertaking, particularly in submarines which are, by their nature, vessels of very high order of complexity.

1.10   Having set the scene with the foregoing account in general terms of how submarine design relates to engineering design at large, we go on to deal in turn with each of a series of technical considerations specific to submarine design before, at the end of the book, returning to the submarine design process. We adopt this approach, despite its apparent interruption of continuity, because we think it the better solution to the presentational dilemma in treatment of the design process, which is whether to describe the process before or after addressing the variety of technical considerations which importantly influence the process itself and the associated explanation of the way in which submarines are configured.

2 MILESTONES IN SUBMARINE HISTORY

TERMINOLOGY: SUBMERSIBLE AND SUBMARINE

2.1   There are fashions in the terms used to describe vessels capable of operating underwater, which are particularly evident when their history is under review. Some are well-established, like the preference for calling these vessels 'boats' rather than 'ships' even when – as applies to ballistic missile deploying submarines – their displacements are some tens of thousands of tons. Others, like the differentiation sometimes made between submersibles and submarines, are contentious and can be confusing. The argument for differentiating is that it was not until the advent of nuclear propulsion, and the associated atmospheric control capability enabling a boat to operate entirely submerged throughout a patrol of several months duration, that the 'true' submarine had arrived. The complementary picture of the submersible is that it describes a boat obliged to operate mainly on or near the sea surface – in order to have access to the atmosphere for oxygen for breathing and for combustion propulsion engines – and which submerges periodically when on patrol for the purposes of concealment, undertaking an attack with torpedoes or avoiding attack on itself.

Our preference is to use the term 'submarine' and we do so throughout this book with its primary focus on naval purposes. We prefer to leave use of the term submersible to commercial circumstances – if that is the wish of the workers in that field – in which it might more closely convey the modes of operation in use there. The fact remains that all submarine boats are submersible – used adjectivally – and to imply a sharp differentiation is misleading.

Throughout the ages, many ingenious devices for going underwater were tried (Table 2.1). Bushnell's *Turtle* 1776 and Fulton's *Nautilus* 1800 were attempts to beat the British blockade of ports by manned submersibles. Most of these earlier designs suffered from the lack of a suitable power plant and it was the development of the battery and internal combustion engine which began to make the submarine a reality as a practical vessel. Many different designs and configurations were tried, by inventors in different countries, that may lay claim to being 'the first'. In

Table 2.1 *Submarine History*

Date	Event/Name
415 BC	Free Divers – Syracuse
330 BC	Diving Bell – Aristotle
1620 AD	Submersible – Van Drebble
	S/M Galley
1776 AD	*Turtle* by Bushnell – USA
1800 AD	*Nautilus* by Fulton
1863 AD	*Plongeur* by Bourgois
1864 AD	*David* – US Civil War Sank
	USS *Housatonic*
1880 AD	Nordenfelt, Goubet,
	Narval by Laubeauf, Holland.
1901 AD	*Holland* No. 1.

this book, we are concentrating on design rather than history of development, but suggest that a designer, however experienced, would do well to study the early ideas. It may well be that what was impractical then has now become possible with modern technology and materials.

In our opinion, two designs at the turn of the century stand out as the fore-runners of the modern submarine. John Holland's USS *Holland* is our choice as the first milestone in the path of development of current submarine designs, because it contained so many of the features now incorporated in those designs. However, because of the practicalities of the day, it was the *Narval* of Laubeuf which proved to be the precursor of most submarines in the first half of the century. His concept was to build the pressure hull within the envelope of a surface running craft, essentially a torpedo boat. As we shall discuss, it was this concept of a surface running vessel capable of occasional submergence which became the practical solution for many decades.

THE FIRST MILESTONE: THE *HOLLAND*

2.2   The *Holland* (Figure 2.1), which was tested by the US Navy in 1899, was the outcome of 25 years of obsessional perseverance of an Irish American schoolmaster, John Holland, with the purpose of producing a practical submarine torpedo boat – initially at least in the hope that it could be used against the British. With his inventiveness, the *Holland* design anticipated to a remarkable degree several features – like low length/diameter ratio, axisymmetric circular form, single screw propeller and small superstructure – shown many decades subsequently to be the most effective ways of configuring a submarine.

John Holland became chief engineer of the Electric Boat Co. of Groton, Connecticut, in the USA, which in 1900 produced a class of six larger boats based on the *Holland* design. The potential of these submarines came to be recognised in the UK – albeit in the face of much reluctance in some parts of the Royal Navy – and, following tortuous negotiations, the British shipbuilders, Vickers of Barrow-in-Furness, in 1901 built a *Holland* type submarine under licence from the Electric Boat Co., known later as *Holland I*. This was not only the first of a long line of submarines built by Vickers, but also the start of a special relationship between the two firms which continues to this day.

On trials, *Holland I* was found to be particularly prone to swamping due to the low freeboard and the necessity to keep open the upper hatch of the low conning tower to provide air for the engine. Vickers were quick to modify the design, making it larger with more freeboard and a higher bridge fin structure, which made running on the surface less hazardous.

In this way, the first class of British submarines, the A Class, started. John Holland's work, as the last individual innovator in the submarine field, was soon left behind by the swift pace at which most of the navies of the world took up this new weapon of war. His contribution to that development was outstanding in the way he devised good practical solutions to the many problems which had, prior to his time, blocked the path to the introduction of a viable submarine.

2.3 In the relatively brief period between the turn of the century and the outbreak of World War I, submarine design progressed so rapidly that the USA, France, the UK – and, significantly, Germany – all had by 1914 substantial numbers of ocean-going submarines. As far as the UK was concerned, although there remained doubt as to the role of the submarine in a battleship dominated navy, there was acceptance of the

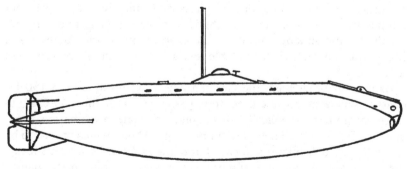

Fig. 2.1   *Holland*

Surfaced displacement	63.3 Tons	Length	53.0 ft	Surfaced speed	6 Kts
Submerged displacement	74 Tons	Beam	10.25 ft	Submerged speed	5 Kts

importance of finding out what they were capable of so that counter-tactics could be contrived. So successor classes to the A Class – designated B and C Classes – had been built at Vickers well before 1914. However, these boats served increasingly to demonstrate shortcomings of the design, ranging from the hazards of the petrol engine for propulsion on the surface to the limitations imposed by the penalties of locating the main ballast tanks inside the pressure hull.

The D Class of submarines for the Royal Navy, the first of which entered service in 1909, effectively broke the link with the *Holland* design. Vickers diesel engines were used in place of petrol engines and main ballast tanks external to the pressure hull were introduced (the *Narval* concept). These boats, and the succeeding E Class, were capable long-range submarines representing an answer to a question which still had not at that stage been answered sufficiently clearly: what were submarines really for?

THE SECOND MILESTONE: THE U35 CLASS U-BOATS

2.4 The German Navy in WWI provided an even more convincing answer to that question. Despite German submarine successes in sinking warships of the Royal Navy in the closing months of 1914, the importance to the British of trade routes led to the decision at the start of 1915 to mount a campaign of unrestricted attacks by U-boats on any merchant ships bound for the UK. The purpose of the campaign was to starve Britain into submission by cutting off the considerable flow of supplies transported to this country by sea.

At the start of the campaign, the German navy had relatively few – 20 or so – small U-boats to deploy. These mostly sank the merchant ships by gunfire with the submarine surfaced, torpedoes being conserved for large targets. Anti-Submarine Warfare (ASW) was non-existent to start with, and the convoy system – of herding merchant ships together to steam within protection afforded by escorting small warships – was not in operation. So, increasingly as the U-boat fleet grew in numbers and capability, the sinkings of Allied merchant ships in the Atlantic and Mediterranean reached formidable levels of hundreds of thousands of tons a month.

As the war progressed, the German navy built up to around 150 U-boats in service at any one time from which we choose the U35 Class to represent (rather symbolically) the second milestone in submarine development (Figure 2.2). They were simple, rugged boats which proved to be very reliable – a distinct step forward from the *Holland*. Characteristically, the design was the outcome of putting emphasis on surface performance, as the boats had to transit using diesel propulsion on the surface and submerged only reluctantly because of the low endurance with battery propulsion. Thus, as can be seen from the illustration, the design incorporated a ship-shape bow, quite large bridge fin (sail) and superstructure, underslung twin screw propellers, a gun, many excrescences and innumerable flooding and venting holes for rapid surfacing and submerging – all of which contributed to very high drag when submerged.

Towards the end of the war the Allies – now including the USA – brought the U-boat threat under some degree of control, mainly by the introduction of the convoy system. Convoys did not lead to a dramatic increase in the numbers of U-boats sunk, because anti-submarine methods were still crude, but the convoy system did reduce the opportunities for U-boats to make successful attacks on merchant ships.

At the end of World War I, the overall balance sheet demonstrated the benefit to Germany of their U-boat fleet: over 11 million tons of Allied shipping sunk and hundreds of warships of all sorts tied down by anti-submarine activities; the cost to the Germans was the loss of around 200 U-boats, which were cheap to build and operate and with a relatively small crew of 40 men.

In little more than a dozen years from the time the *Holland* first went on trials the submarine had won its spurs, though not in a knightly way. Allied and German submarines were also used to good effect in attacks on surface warships and the sinkings were quite out of proportion to the number of submarines applied to that purpose, but the major impact was that of the U-boats on Allied merchant ships and the curtailment of imports to Europe.

Between the two World Wars other roles and capabilities were explored for the submarine. In the UK the large K Class – which actually emerged in World War I – was tried, a vessel with a surface speed capable of keeping up with the fleet. The hazards of steam boilers, requiring funnels to be shut down when submerging, became sadly all too apparent. Similar ill fate met the M Class with one vessel mounting a large calibre, 12 inch, gun and another housing a small aircraft. The French *Surcouf*, with its heavy armament of a twin gun turret, survived into World War II.

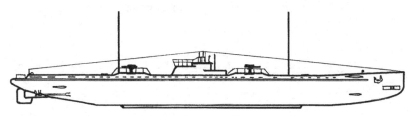

Fig. 2.2   U35

| Surfaced displacement | 685 m³ | Length | 64.7 m | Surfaced speed | 16.4 Kts |
| Submerged displacement | 844 m³ | Beam | 6.32 m | Submerged speed | 9.7 Kts |

THE THIRD MILESTONE: THE TYPE XXI U-BOATS

2.5 Both Britain and Germany entered World War II with the lessons of the First World War clearly in mind. Not only did Britain realise from the start that there would be unrestrained submarine attacks on merchant shipping bound for the UK and that convoys should be introduced, but it was also able right away to deploy an effective underwater detection system, ASDIC, mounted in the ships escorting the convoys. ASDIC, an acronym for an active sonar system, had been developed in the 20 year interval between the wars and in conjunction with depth charges was believed to be the answer to the submarine threat.

Germany for its part had developed the 'wolf pack' method of attack on convoys, in which a group of U-boats was directed from shore to a convoy and attacked on the surface at night from several different directions, confusing both the merchant ships and their escorts. However, it took some time for the German navy to build up sufficient numbers of U-boats to support that approach, and so it was a full year until the onslaught on merchant shipping in the Atlantic got under way in mid 1940. Once under way it was very effective because the U-boats were deployed on the surface when attacking – usually with torpedoes – so that ASDIC was useless (it depended on the submarine being submerged) and the U-boats were fast enough on the surface to make their getaway in the confusion. At that stage in the war, in fact, U-boats spent most of their time on the surface, for transiting at night as well as attacking, and submerged only during the day for concealment or lurking close to convoys waiting for nightfall.

The U-boats in service at that stage were much like their predecessors of the previous war in the characteristic emphasis on surface performance, though the Type VIIC was a very well-developed version, far better than its British equivalents (Figure 2.3) It was this need to operate so much on the surface that led to the development of a successful counter to the 'wolf packs': radar deployed by patrol aircraft. This enabled aircraft to attack U-boats on the surface at night until the German navy introduced a radar-warning device mounted on the bridge which gave a submarine, subject to imminent attack, enough time to crash-dive and escape.

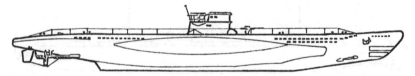

Fig. 2.3   Type VIIC
Surfaced displacement	762 m³	Length	67.10 m	Surfaced speed	17 Kts
Submerged displacement	871 m³	Beam	6.23 m	Submerged speed	7.6 Kts

By late 1942, Germany had over 300 U-boats in commission and even allowing for the fact that only about one in three was ready for sea service (the others either being refitted, on trials or training) and half of those were on passage, around 50 U-boats were available for attacking ships of all sorts – and perhaps 30 or so would be in the Atlantic attacking merchant shipping with devastating effect.

Gradually the efficacy of radar was improved and made less readily detected; the range of marine patrol aircraft was further extended and the number of aircraft employed on anti-submarine work greatly increased; ASDIC was also made more effective and fitted to more warships, which could be used not just as escorts to convoys but also actively to hunt 'wolf packs' on their way to intercept convoys. By the spring of 1943 the German navy was forced to recognise that these measures were cumulatively so telling that the U-boat campaign had to be abandoned and the Battle of the Atlantic was all but over.

2.6 One measure then introduced in an attempt to regain the lost ground was to equip the U-boats with a schnorchel, a device which enabled the submarine to use its diesel engines while proceeding submerged at periscope depth by drawing air down through a non-return valve mounted at the top of a tubular mast. But use of the schnorchel reduced transit speeds from up to 16 knots on the surface to about 5 knots at periscope depth, virtually trebling passage time from base to patrol area. Thus although U-boat attacks on Allied shipping in the Atlantic were resumed in late 1943, sinkings were greatly reduced and achieved at a severe toll in U-boat losses, due to improved Anti-Submarine Warfare (ASW) tactics employed by surface warships.

Another approach to improvement in U-boat capability, which accepted that emphasis had to be switched entirely from surface to submerged performance, was represented by the Type XXI submarine, and this we select as our third milestone in submarine development. (Figure 2.4) In that design, priority was given to reducing underwater drag, both by drastically cutting down on appendages and excrescences and by streamlining the hull. Battery size and fuel stowage were increased and underwater speed was raised to 18 knots (double that in previous U-boats) and range much improved.

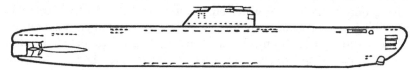

Fig. 2.4   Type XXI

Surfaced displacement	1602 m³	Length	76.7 m	Surfaced speed	15.5 Kts
Submerged displacement	1800 m³	Beam	6.61 m	Submerged speed	17 Kts

The Type XXI submarine was a major change of direction in submarine design which came to be subsequently adopted by all navies because it was the proper response to inexorable progress in ASW techniques. The German navy also attempted the even more revolutionary innovation in submarine technology represented by Walther steam turbine propulsion machinery using concentrated hydrogen peroxide as the source of oxygen, which would have given even higher underwater speed for several hours, but we are in no doubt that the Type XXI submarine was the more sensible step to have taken.

Although the first of class U-boat to the Type XXI design, U2311, did go on patrol in April 1945, it could not be used offensively before the war in Europe was over. Contrast of the balance sheet with World War I is instructive: nearly 1200 U-boats were built in the later war, which sank 14 million tons of shipping and tied up many hundreds of warships and thousands of marine patrol aircraft; the cost to Germany of nearly 800 submarines lost was much higher than in the earlier war, but ultimately still not a high price to pay for so much damage to the Allies. In retrospect, the outcome of the Battle of the Atlantic was inevitable, though while it was in progress it appeared to be a very close run thing for Britain.

2.7   We have again focused the foregoing account of submarine warfare in World War II, as for World War I, on attacks on merchant shipping and on the role of the U-boats in that mode. As previously, submarines of the various navies involved were also used to telling effect in attacks on surface warships of all kinds. In particular, the US Navy employed their fleet submarines against the Japanese Navy and merchant shipping to great effect. The design of these submarines, which was developed well before the war, achieved the long ranges and high speeds necessary in the Pacific theatre, still retaining emphasis on surface performance.

We should also mention that while in World War II the gun gave way to the torpedo as the primary means of attack by U-boats on merchant ships, a considerable number of sinkings was also caused by mines laid by submarines in the approaches to ports and harbours.

The Royal Navy had entered World War II with similar submarines to other nations, the S, T and U Classes. Later the A Class was developed, still with the principle of being able to make long surface transits and having only limited underwater endurance. At the end of the war the remaining T and A boats were fitted with the snort system, an adaptation of the schnorchel, and some T Class were lengthened to include greater battery capacity and higher submerged performance. The subsequent class of *Porpoise* and *Oberon* submarines followed the same style of design, incorporating many features of the Type XXI. With the emphasis on longer submerged operations the primary detection method became acoustic in the form of sonar, particularly in passive or listening modes, and considerable emphasis was placed on the problem of producing a 'quiet' submarine. Thus the *Porpoise* and *Oberon* boats, though not particularly innovative in configuration, represented a major advance in the reduction of radiated noise, allowing an operating mode which rendered them virtually undetectable to passive sonar.

THE FOURTH MILESTONE: THE *ALBACORE*

2.8   In the early 1950s initiative in submarine development passed to the USA. Directly after the end of the war the US Navy converted the fleet submarines to *Guppies* for better underwater performance, based on the German Type XXI approach of streamlining, reducing appendages and increasing battery power, and went on to design a new class of battery-drive submarines, the *Tang* class, following the same pattern. Then, starting in 1948, came the momentous step to nuclear propulsion in the *Nautilus*, which was in other respects relatively conventional in form and arrangement as a deliberate act of policy. At about the same time the US Navy commenced research work at what was then the David Taylor Model Basin on a radically new shape of submarine, leading to the design and construction of an experimental battery-drive submarine, the *Albacore*. It is the *Albacore* rather than the *Nautilus* that we have chosen to be the fourth milestone as – for reasons which will become apparent – we reserve the milestone marking the great contribution of nuclear propulsion in submarine development to a later US Navy design.

Not a lot has been published about the *Albacore* except that Arentzen and Mandel give a full account of how the tear-drop form contributed to the primary goal of the design: achievement of high propulsive efficiency to maximise submerged performance at the expense of surface capabilities. Figure 2.5 helps to indicate how that goal was attained, and show how fully John Holland anticipated the measures required for the purpose: an axisymmetric (solid of revolution) form of relatively small length/diameter ratio (seven to one for the *Albacore* compared with just under six to one for the *Holland*); a large diameter screw propeller on the

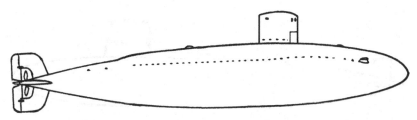

Fig. 2.5   AGSS *Albacore*

Surfaced displacement	1500 Tons	Length	210.5 ft	Surfaced speed	25 Kts
Submerged displacement	1850 Tons	Beam	27.5 ft	Submerged speed	33 Kts

axis running at low r.p.m.; a small bridge fin and superstructure; and a drastic reduction in the number and size of excrescences, appendages and holes in external structure. Many of those features can be recognised as particular to submarines, but it can be inferred that the original interest in the blunt-nosed and fine-tailed form (the tear-drop shape) stemmed from earlier aerodynamic research on airships, a vehicle which the submerged submarine closely resembles.

The parallel paths of development represented by the *Nautilus* and *Albacore* showed a high degree of imagination as well as courage on the part of the US Navy. The main purpose of *Nautilus* was to demonstrate how the Pressurised Water Reactor (PWR) stood up to service at sea, while the main purpose of *Albacore* was to demonstrate that a submarine of unique shape (with manifest disadvantages for surface performance) could nevertheless be operationally acceptable through its greatly improved submerged performance, including handling characteristics of good course and depth manoeuvring and control. Both submarines were capable of relatively high speeds, the *Nautilus* for prolonged periods and the *Albacore* for short bursts (long enough, though, for test purposes). Both were outstandingly successful, encouraging the US Navy submarine designers to seek another goal – that made possible by combining the innovations of both developmental submarines – which we now turn to.

THE FIFTH MILESTONE: THE *SKIPJACK* CLASS

2.9   It is the *Skipjack* Class (Figure 2.6) that we nominate as the fifth milestone, for it was an extraordinary achievement – following so closely on the heels of the *Nautilus* and *Albacore* – to combine the PWR nuclear propulsion plant with a hull of tear-drop shape, and for both to be incorporated in a fully operational attack submarine, the *Skipjack*, which entered service in 1958 as the first of the SSN585 Class, just four years after design commenced. Amongst the many notable successes in the design was a substantial reduction in displacement from that of *Nautilus*, while still developing the same propulsive power, giving an even higher underwater speed as well as greater manoeuvrability, a formidable combination.

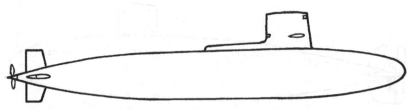

Fig. 2.6   USS *Skipjack*
Surfaced displacement	3075 Tons	Length	251.8 ft	Surfaced speed	16 Kts
Submerged displacement	3513 Tons	Beam	31.6 ft	Submerged speed	30+ Kts

For the Royal Navy, the *Skipjack* development was momentous for another reason: the US Government of the day made available to the UK a PWR propulsion plant, primary and secondary machinery, as fitted in the *Skipjack* Class, for incorporation in the RN's first nuclear submarine, the *Dreadnought*. This far-sighted gesture on the part of the USA, a very material manifestation of the special relationship between the two countries, enabled Britain to make the transition from conventional to nuclear submarines much sooner than would otherwise have been possible. Moreover, it brought back together again, some 50 years later on, the two shipbuilders who had first co-operated when the Electric Boat Company enabled Vickers Shipbuilders to build the *Holland I* under licence. As recorded in *The Building of the Two Dreadnoughts*, (Baker and Rydill, 1983) the benefit to Vickers and the RN extended beyond the boundaries of the nuclear propulsion plant into the new technology and research findings identified by the US Navy to be needed for a sound understanding of the design problems associated with high speed submerged operation. Actually, the virtues of the *Albacore* form apply equally to conventional submarines, even though with batteries attainable speeds are not as high as with nuclear plant nor sustainable for very long. In fact, the US Navy did introduce a class of diesel electric drive submarines, the *Barbel*/SS580 class, almost concurrently with *Skipjack* Class. The *Barbels* were entirely successful in their own right but, because the US Navy became convinced that the right way to go in subsequent development was with nuclear propulsion plant only, there was no further US investment in conventional submarines after the *Barbel* Class. It remained for the Dutch Navy to first take up the application of the *Albacore* form to conventional submarines, with the *Swordfish* design. Then the Royal Navy, when at last it had to replace the very successful *Porpoise* and *Oberon* Class submarines, pursued the *Albacore* approach in the largely Vickers designed *Upholder* Class.

2.10   The design of military submarines in the ensuing thirty years has not seen any major changes in the innovative features of design represented by our selected examples. However, progress has been made along three separate avenues: size, speed and silence.

The submarine had in the past been a relatively small vessel of no greater than a corvette or frigate displacement and with a small crew. The advent of nuclear propulsion resulted in a twofold increase in displacement, approaching that of a light cruiser, and such submarines have become a major element of the fleet. The introduction of the underwater launched ballistic missile system resulted in another dramatic increase in the size of submarines, of which the Russian Navy's *Typhoon* Class is the outstanding example. Thus submarines in terms of displacement have become some of the largest units of the major navies, bigger than many battleships of the early part of the century and only exceeded by major aircraft carriers in the modern fleets. Such vessels are indeed a major undertaking in design and construction.

The low drag *Albacore* form combined with nuclear power has given the submarine a speed capability underwater in excess of most surface displacement vessels. The reported performance of the Russian *Alpha* Class puts it to the fore in sheer speed capability. The modern nuclear submarine no longer has to wait in the path of targets but can close or overtake them. However, this entails a high demand for power which can only be sustained for long periods with the energy from a nuclear reactor. The search for a more economic power plant and reduction in the power required for speed is one of the areas of future development, where some indications of progress are appearing.

With many navies going 'underwater' to a significant degree, most above water sensors and surveillance systems are by-passed and the primary method of detection, particularly between opposing submarines, is by means of acoustics. The preferred mode is passive, i.e. listening to sound emanating from other submarines; the alternative active mode, in which the water is ensonified and the reflected sound waves are picked up, has the disadvantage of revealing the hunter to the hunted.

As mentioned earlier, later development of the conventional diesel electric submarine has introduced noise reduction features so that, by adopting a quiet mode running slowly on battery/electric motor propulsion with virtually all other systems stopped, the submarine radiates hardly any noise that can be detected by other vessels, whether surface ships or submarines, while at the same time it is able to pick up even very low source strength noises from those other vessels.

In that sense the advent of nuclear power, whilst conferring many considerable advantages, was a retrograde step in noise reduction as it involved a complex, high-powered plant which could not readily be shut down. Consequently, although the configuration of the modern nuclear attack submarine has not changed greatly from the *Skipjack* innovations, considerable advances have taken place internally with the purpose of restoring the concept of a quiet submarine capable of avoiding detection.

REVIEW

2.11   The path of the development of the submarine from the *Holland* to the *Skipjack* took about 60 years and was in some regards almost circular. That, however, was due more to the foresight of John Holland than to a tendency for history to repeat itself. It seems likely that he really did envisage his submarine as a vessel intended to operate submerged, surfacing at the beginning and end of a patrol and that is what the modern submarine, whether conventional or nuclear in propulsion, can now do. However, in the interim, through two world wars, the submarine was obliged to operate mostly on the surface – submergence was an essential capability, but one more likely to be invoked out of necessity than preference. What has transformed that situation in the last 30 years or so has been a series of technological developments – of which nuclear propulsion is unquestionably the outstanding one, but not the only one. The snort system is also significant, as are the means of improved atmosphere control inside the submarine. Further technological improvements are still sought, particularly airless propulsion submerged to extend the underwater endurance of the conventional submarine before the indiscretion of snorting becomes necessary. All these topics we return to later in the book. What is not in doubt is the cost effectiveness of the submarine as a weapon of war, no longer just as a patrol or attack vessel but also as a platform for ballistic missiles. As things now stand the clear conclusion is that the biggest threat to a submarine is another submarine – and this applies even though there have been great strides in ASW from surface warships with helicopters and from marine patrol aircraft.

This brief survey of historical milestones has placed emphasis on military submarines, partly because our book is directed towards the design of major manned submarine vehicles which are at present all military. The other reason is that most of the main technical advances have been driven by real or potential military conflict underwater and the investment in the necessary research and development has been funded by Government Defence sources.

Nevertheless, there is a small but growing interest in underwater craft in the civil field. These are for exploration, oceanographic survey, salvage, mineral and oil recovery and even as leisure craft where the water conditions allow viewing. A large number of craft, some of strange configuration usually designed for a specific purpose, now exist throughout the world. Whilst no survey of the development of such craft has been attempted in this book, the principles of submarine design which are described in ensuing chapters apply to all such craft, though the emphasis of different features may differ. Thus some civil applications involve no onboard manning, e.g. Remote Operated Vehicles (ROVs), and the requirement for speed is usually low, which permits the design of much smaller craft and possible reduction of safety features. On the other hand, the vehicles may have to perform quite complex tasks underwater, involving greater control of attitude and position. Certainly many such craft have a requirement for greater depth of operation than military vehicles, so structural considerations and weight of the pressure hull dominate the design.

It may have been noticed in the examples of technical progress that no emphasis was placed on the achievement of greater diving depth. In fact, there has been a progression in the diving depth capability of the submarine with improved methods of analysis for structural efficiency, higher grade steels and alternative materials and the associated capability to construct stronger larger hulls. It has nevertheless proved difficult to single out any particular design where this progression represented the significant advance in technology. Thus, for example, the design of *Dreadnought,* in which both authors were involved, was of a vessel considerably larger, faster and deeper diving than any previous design for the Royal Navy, which entailed the solution of many detailed design problems which hitherto had not been addressed. Subsequent designs have been of even greater depth, but, as discussed in the Structural Design chapter, there remains cause for debate on the value of great diving depth for the military submarine. Very deep diving has been progressed more in the civil field for exploration and salvage in the deepest areas of the oceans.

3 SUBMARINE HYDROSTATICS

FIRST PRINCIPLES OF FLOTATION

3.1 To naval architects the hydrostatic properties of vessels floating on the water surface represent a fundamental part of their stock in trade, and that familiarity readily reads across to submarines on the surface; the hydrostatic properties of submerged submarines are less familiar to naval architects in general, but they can identify the parallels without difficulty. For most other engineers the subject of hydrostatics may not be so familiar. For that reason Appendix 1 is provided which gives the broad principles on which the following discussion is based.

The submarine, like any other marine vehicle, has to be designed to float in the water where its weight is supported by the buoyancy forces due to the displacement of water by its hull. For a surface ship the objective is not simply that it should float at the surface but also that it should remain afloat even if some damage resulted in part of the hull being flooded. To cater for this means the provision of a substantial watertight volume of hull above the normal waterline, (Figure 3.1). This enables the surface ship to accommodate changes in weight by adjusting its level of flotation, i.e. by slightly sinking or rising. The amount of sinkage is governed by the area of the hull at the waterline (the Waterplane Area).

As well as just floating, the ship must remain upright in calm water. This involves the consideration of transverse stability and the concept of GM (Metacentric Height) and GZ which, as shown in the Appendix, is also governed by the characteristics of the waterplane area. (Figure 3.2(a)).

3.2 A submarine on the surface has to satisfy the same hydrostatic principles. However, as will be discussed, it has a small watertight volume above water and relatively low waterplane characteristics. The condition is therefore more critical than for a surface ship. It can accommodate very little weight change and can sustain only a small amount of flood damage.

The act of submerging, which the surface ship designer tries to avoid, involves the deliberate flooding of part of the surfaced buoyant volume. The extent of flooding has to be constrained so that, when submerged, the remaining total of buoyant volume just supports the weight of the vessel. This represents an entirely different hydrostatic condition to that of a surface ship. It will be seen that it requires the provision of special tankage both to bring about the transition from surfaced to submerged and to sustain the balance between weight and buoyancy whilst submerged. The absence of any reserve of buoyancy whilst submerged leads

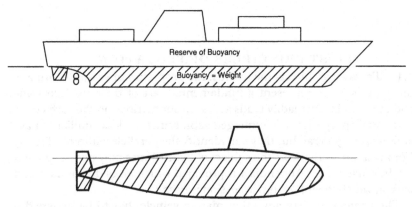

Fig. 3.1   Surfaced ship – Surface submarine

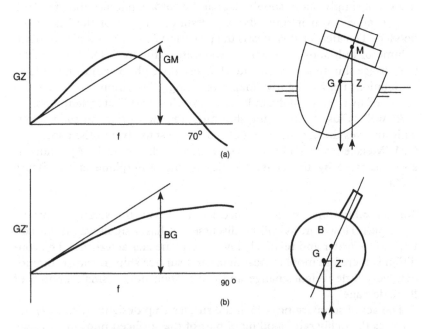

Fig. 3.2   Heeled stability graph surfaced and submerged

to a direct relationship between weight and total hull volume and introduces a constraint which permeates the whole of the submarine design. With no waterplane area stability relies on G being below B (Figure 3.2(b))

This chapter now goes on to study these features in more detail to provide a rationale and basis for sizing the associated tankage.

SUBMARINES ON THE SURFACE

3.3   It will be apparent from Chapter 2 that the modern submarine, whatever its propulsion plant, cannot afford to operate for other than short periods on the surface because of the ease with which it can be detected and attacked. Nevertheless, it has to be on the surface when leaving and returning to its base and, when alongside in harbour, hatches have to be open for access and storing. Thus design has to make provision for a safe surface condition, even though priority is given to getting the best achievable performance submerged.

The primary design requirement to be met in the surface condition is that part of the watertight containment (the Pressure Hull) should be above the calm water level. Exactly by how much is not fixed, but neither is it completely arbitrary. As guidance, there is the requirement for sufficient freeboard to the pressure hull hatches to avoid accidental flooding when they are open, and the achievement of a sufficiently shallow draught for the submarine to negotiate restricted waters which are common at harbour entrances. There might also be a case for having more than one surface condition so that the submarine could be at lower buoyancy preparatory to diving than when actually entering and leaving harbour; in the former circumstances, pressure hull hatches would be shut and probably awash – only the hatch at the top of the conning tower would be open.

Whatever the surface condition, there will be a proportion of the hull volume above the waterline at which the submarine is to float. When the submarine submerges, the hitherto above water volume will now displace water, and thus to change from surfaced to submerged the submarine must either increase its weight or reduce its buoyancy; to surface from submerged the opposite must happen, that is, either an increase in buoyancy or a reduction in weight. For small submersibles the transition can be effected by a buoyancy change using flexible bags on a flotation collar, which for surfacing is inflated by gas or air pressure and for submerging is collapsed by venting. However, to date no suitable flexible material for the purpose has been identified which would be sufficiently robust for application in a larger submarine, though the quest is worthwhile as the approach has the attraction of minimising the underwater volume which has to be propelled by the often limited stored energy on board the submarine. Thus as things stand at present the submarine designer has to accept an essentially fixed and integral structure within which to bring about the transition between surfaced and submerged conditions.

Referring to Figure 3.3, if we regard the outer surface or envelope of the submarine to be its boundary with the sea, then the volume enclosed in this boundary must be sufficient to enable the submarine when surfaced to float with some of this volume above the surface. To submerge, the submarine has to be able to increase its weight until it equals the buoyancy of the total envelope volume. Since the only readily available source of weight is the surrounding sea, the natural thing to do is to flood part of the space within the outer envelope with sea water; to surface this water is then discharged to sea. This space constitutes volumes of the submarine which are termed the Main Ballast Tanks (MBT). Other spaces within the outer envelope may be flooded, either wholly or partially, depending on where they are located relative to the waterline. Such volumes are termed Free Flood (FF) Spaces and do not contribute to the buoyancy of the hull.

This process, which is used by virtually all submarines, introduces one of the classical dilemmas of naval architecture when dealing with the flooding of surface ships: whether to treat this flooded water as added weight in the submarine or the volume occupied by the flooded water as causing lost buoyancy. There is no simple answer as it is sometimes convenient to treat it in one way or the other or even a combination of both; also the concepts of added weight or lost buoyancy can be put to good

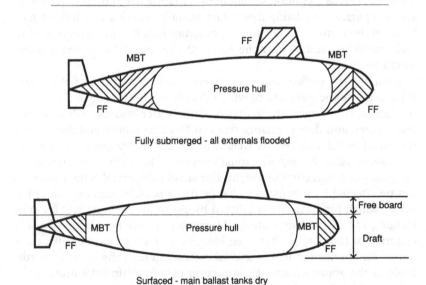

Fig. 3 3   Flooded volumes surfaced and submerged

use to provide a check by independent alternative calculation. What we do recommend is that the concept of an outer envelope which defines the boundary between the sea and the submarine should be retained and reverted to in case of doubt. As we show later the outer boundary concept is also relevant when considering propulsive power and hydrodynamic control as everything within that boundary to all intents and purposes moves with the boat.

As well as designing a submarine to have positive surface flotation at a reasonably level trim there is also the need to satisfy the heel stability criterion by having a positive GM. As we explain in more detail later, the configuration of a submarine makes it difficult to provide the vessel with large heel stability. Motion on the surface is usually characterised by slow heavy rolling, sometimes to alarming angles, whereas, fully submerged, little motion occurs. Hence, the surface is not the place for a submarine in heavy weather..

ARRANGEMENTS OF MAIN BALLAST TANKS

3.4   As we described in Chapter 2 and return to in Chapter 6, the US Navy's research and development into the best form of a submarine for efficient propulsion and good manoeuvring and control submerged, culminating in the *Albacore* trials, showed it to be a streamlined (tear-drop) body of revolution with a blunt rounded nose and tapering conical tail. Within that outer envelope there will be a pressure hull which for structural efficiency reasons, as we discuss in Chapter 5, is best configured as a cylindrical body of nearly constant diameter circular sections with domed ends. This arrangement is shown in Figure 3.4(b), together with some other configurations encountered in submarine practice and which we also refer to from time to time.

The differences between the shapes of pressure hull and outer envelope result in spaces between them, and in design one aim could be to minimise their extent so as to achieve the most volume efficient form, i.e. one in which total volume is made as small as possible in order to reduce the power to propel the submarine. However, that aim can lead to conflict with that of simplicity and efficiency of pressure hull structure, e.g. the waisted single hull configuration of the US Navy's *Skipjack* class, (Figure 3.4(c)). There may be other priorities, for example, the improved resistance to damage represented by the double hull arrangement, which provides a complete wraparound outer envelope to the pressure hull; this could actually give an excess of usable space between the two boundaries, even in diesel-electric drive submarines in which diesel fuel is usually stowed externally.

Not all the space between pressure hull and outer envelope can or need be utilised for floodable tankage. Some of it is occupied by various external equipments including sonars, torpedo tubes, hydroplane and

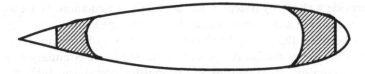

(a) Exposed pressure hull shaped to form

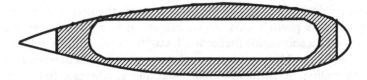

(b) Enclosed cylindrical pressure hull

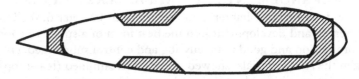

(c) Waisted pressure hull partially exposed

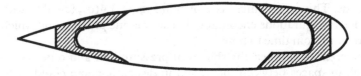

(d) Exposed pressure hull reduced at ends

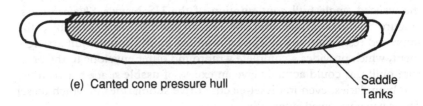

(e) Canted cone pressure hull

Saddle
Tanks

Fig. 3.4   Main ballast tank arrangements

rudder gear, anchor, etc. Some of it may not be capable of useful application at all and necessarily has to be free flooding. When these competing functions are taken into account, it will be seen that the amount of floodable tankage may be quite limited (other than in double hull configurations, which can err in the other direction). It is possible to augment external floodable tanks with floodable tanks inside the pressure hull, all with the purpose of enabling the submarine on the surface to submerge, and vice versa. We discuss the advantages and disadvantages of such an arrangement later.

The most useful floodable tankage is that around the lower part of the pressure hull. There is little value in locating some floodable tankage above the pressure hull as it would be exposed when the submarine is surfaced and so cease to contribute to buoyancy, thereby failing to increase pressure hull freeboard. It is for these reasons that the superstructure casing, which provides a walking deck, and the bridge fin (sail) do not normally contain tankage and are made free flood spaces, i.e. when the submarine submerges they flood up and when the submarine surfaces they drain down, care being taken to minimise any tendency to air pocketing. The extreme bow and stern are also free flooding, but as they remain below the surface they are permanently flooded except when the submarine is in dock.

We can now proceed to discuss how this group of floodable tanks, termed MBTs, can be sized and located in the design process.

3.5 Main Ballast Tanks are subdivided longitudinally by transverse bulkheads and may also be divided port and starboard. Transverse subdivision is provided partly by natural structural barriers but also deliberately to reduce vulnerability, so that if the submarine sustained rupture to the outer shell it should still be able to surface on the tanks remaining intact. Another purpose of transverse subdivision is to enable the tanks to be used differentially to alter the surface condition. Thus if it were required that the submarine should proceed on the surface but ready to dive quickly, it could run in a low buoyancy condition with some of the tanks flooded; alternatively, it might on occasions be advantageous to run on the surface with heavy stern trim so as to increase the immersion of the propulsor, which can be achieved by leaving the after tank flooded.

It is usual to associate the capacity of the MBTs with the factor termed reserve of buoyancy (ROB), which in submarines is the ratio of the effective volume of these tanks to the volumetric displacement of the submarine on the surface. The effective volume of the tanks is their total blowable volume, i.e. total volume less that of the residual water left at the bottom of the tanks. The surfaced displacement is, like any other surface ship, essentially the weight of the submarine excluding free flood water. As discussed later, in the submerged condition, this weight is primarily supported by the buoyancy of the fully submerged pressure hull.

The amount of the reserve of buoyancy incorporated in a submarine design depends on its size and on other considerations touched on earlier. Thus it is likely to be larger in small submarines because it is desirable to have a reasonable minimum freeboard. The reserve of buoyancy ratio can vary between as little as 10% and as much as 20% (may be even more in double hull configurations). These values are to be compared with upwards of 100% in surface ships, depending on their cargo density; but then those ships are not intended to submerge. Where internal MBTs are provided, they may serve to increase a low reserve of buoyancy on external MBTs alone.

It follows from the foregoing definitions that in most submarines the surface displacement is largely dictated by the volumetric displacement of the pressure hull, the external equipment being only a small percentage of this major item. Thus, as a first approximation in initial design, the volume of the MBTs can be determined by:

$$MBT \ vol \approx PH \ vol \times ROB$$

3.6 The longitudinal disposition of MBTs is another important consideration in design. If the submarine is in balance and with level keel when submerged then to achieve level keel on the surface the centre of volume of MBTs must be close to the longitudinal centre of buoyancy of the submerged submarine. This might be difficult to achieve if the spaces at the ends of the submarine were the only places for MBTs, but since it would be unacceptable to have either a large bow or stern trim angle on the surface with all MBTs blown, another solution would have to be found – either by reducing the bow or stern tankage or by introducing internal MBTs if a suitable location could be found.

Turning to the issues which bear on the choice between internal and external MBTs, it is inefficient to have these tanks inside the pressure hull because they would have to be enclosed within heavy structural boundaries capable of withstanding sea pressure. MBTs external to the pressure hull can easily be arranged so that they are equalised with sea pressure at all times, and then they do not require any heavy structure. Instead they can be designed to withstand with an appropriate factor of safety just the air pressure required to blow the water from the tanks through the flood holes at the bottom at the selected rate of discharge – which is a much less demanding structural task.

Sometimes, as we have already indicated, the arrangement of the submarine might necessitate a combination of internal and external MBTs. In that case it is conceivable that sufficient tankage could be arranged externally to provide for safe surfacing in a low buoyancy condition, which could subsequently be increased in slow time when the submarine was on the surface by pumping water from 'soft' internal tanks, i.e. tanks whose structure was incapable of withstanding full sea pressure because they could not possibly be exposed to it. Overall, the combination of internal and external MBTs would have to be suitably located to give nearly enough level keel in harbour. It might help to have a small internal ballast tank located aft in the submarine which would be flooded when the submarine was at low buoyancy on the surface – giving a stern trim for propulsor immersion – and which would be blown to give level keel in full buoyancy; such a tank might also be used in harbour to increase the freeboard at the fore end of the submarine during torpedo loading.

3.7 Another consideration to be taken into account when sizing and locating MBTs is the extent to which they could be used when the submarine is submerged to counter the effects of internal flooding due to an accident or attack leading to penetration of the pressure hull. It has to be recognised that the effectiveness of this counter-blowing is largely dependent on the rate of air blow which could be achieved, in an emergency, to discharge water from the tanks. We discuss this emergency recovery aspect later in the book in Chapter 8. For now we observe that it is necessary for the flood holes at the bottom of the MBTs to be as large as can be contrived to ensure the largest achievable rate of discharge of water from the tanks; this would also aid rapid flooding of the tanks. A problem arises in that large openings in the outer hull can cause noise, vibration and additional drag; measures have been developed for closing the holes while automatically enabling flooding and blowing to take place, or alternatively for fitting a grille across each opening to reduce adverse flow effects.

It is important that the flood holes should be in the lowest regions of the MBTs, for otherwise a larger amount of residual water would be retained than is desirable. (Figure 3.5) For similar reasons the vent

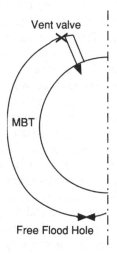

Fig. 3.5   Location of flood and vent holes

valves which allow air to escape from the tanks when submerging are sited at their highest points to ensure that all air is expelled on opening the valves; as we describe later, any trapped air could cause difficulty in controlling depth. It is also preferable for the inlet valves by which high pressure air is blown into the MBTs to be sited high up so that the air discharges into the tops of the tanks and can properly be assumed to stay above water as the water level drops. It is sometimes the practice to provide separate valves to allow low pressure air from a blower to be used to discharge the latter part of the tank water once the submarine has reached the surface and can open up its ventilation snort mast to draw air on board.

Whilst the cross-sectional shape of MBTs is often governed by the concentric circular form of the pressure hull and 'ideal' outer envelope, it is for consideration in design whether to depart from that configuration, particularly in boats which might be required to undertake long surfaced passage. To improve initial stability the waterline beam should be as large as feasible. In some submarines this need has led to the adoption of wall-sided saddle tanks, conferring the higher metacentre characteristic of surface ship shape sections, though the limited reserve of buoyancy means that there would be a rapid fall off in righting moment if the boat were to heel to large angles.

SUBMARINE SUBMERGED

3.8 When a submarine submerges by flooding its MBTs it enters an environment and embarks on modes of operation quite unfamiliar to surface ship operators and to surface ship designers. To start with, the submerged submarine has freedom to manoeuvre in all three dimensions, a capability akin to that of the aeroplane, including the presence of a 'ceiling' and 'floor' which constrain freedom of manoeuvre, though for different reasons. However, unlike the aeroplane, most submarines do not depend on having ahead motion to provide support to their weight by dynamic lift, even though the submarine does invoke hydrodynamic lift for depth and course control purposes when it is under way.

In fact, the closer analogy to the submarine in the aeronautical field is the airship, and once submarines were able to escape from being tied to operating mostly on or close to the water surface, submarine research, like that on *Albacore*, was able to draw on airship research in the 1920s and 1930s. The airship at rest supports the majority of its weight by buoyancy and achieves aerostatic stability in both roll and pitch by keeping its centre of gravity below its centre of buoyancy and in the same vertical; once under way it needs to achieve aerodynamic stability, which it does through after stabiliser surfaces, and flight controllability, which it does through rudders and ailerons; also when it is under way there is a need to minimise its aerodynamic drag, which it does by adopting an

elongated tear-drop shape shown by research to be optimum for the purpose. Practically all these considerations and the research and development findings read across to the submerged submarine, and the parallels
are addressed in later chapters.

3.9 To pursue the hydrostatics of submarines, because they depend on
buoyancy to support their weight they are sensitive to variation in both
buoyancy and weight. Exact equality of the two – which is, for practical
purposes, virtually a hypothetical condition – is termed 'neutral' buoyancy; if weight exceeds buoyancy, known as 'negative' buoyancy, the
vessel will drop unless corrective action is taken either to reduce weight
or increase buoyancy; conversely, if buoyancy exceeds weight, known as
'positive' buoyancy, the vessel will rise unless corrective action is taken
either to reduce buoyancy or increase weight (Figure 3.6). Variation in
buoyancy will mainly be due to change in density of the supporting
medium, seawater, while variation in weight will mainly be due to consumption of fuel, stores or weapons. The state of equilibrium in depth

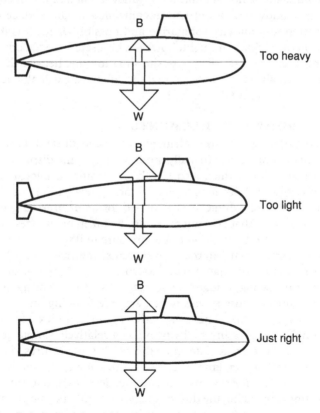

Fig. 3.6   Relative weight and buoyancy

tends to be unstable with the vessel at rest, so that if the vessel were required to 'hover' a special control device would have to be provided. Once the vessel gets under way, the hydrodynamic forces then available from control surfaces enable the relatively small differences between weight and buoyancy normally encountered to be overridden readily enough. Nevertheless, it is sensible to keep weight and buoyancy as nearly equal as can be contrived and means for doing so are provided in the vessel.

3.10 It could be open to debate as to why a submarine should be designed to achieve neutral buoyancy. Why not design for a heavier or lighter weight relative to buoyancy and provide hydrodynamic lift similar to an aircraft? To do so would mean that the submarine would have to sustain speed whilst submerged in order to retain control of depth. The simple answer is that most submarines have an operational requirement which requires them to stop or move at very low speed. If they cannot achieve neutral buoyancy these manoeuvres would not be possible. Another argument is related to safety. If the submarine experiences a power failure and is not in neutral buoyancy it will lose depth control. If lighter than buoyancy it will rise uncontrolled to the surface. This is embarrassing to a military submarine and possibly dangerous if surface ships are in the vicinity. It might, however, be a safety design criterion for a passenger leisure craft. The opposite situation, i.e. heavier than buoyancy, is potentially catastrophic in that the submarine will plunge beyond its collapse depth or hit the sea bed.

BUOYANCY ELEMENTS

3.11 Considering first the elements that constitute buoyancy, the major source of buoyancy (usually around 90%) is the displaced volume of the pressure hull, which also largely determines the location of the centre of buoyancy both longitudinally and vertically. Pressure hull volume and centres of volume are calculated for its unstressed geometry – later it will be seen that account must also be taken of changes in geometry under pressure. In addition to the pressure hull there are elements of buoyancy arising from external components, the main ones being the external structure associated with external tanks, fore and after ends, superstructure casing, bridge fin and control surfaces. For the most part these structures enclose spaces which are free flooding and so displace just their own volume. These are tedious and complex geometries to assess (unlike the pressure hull which has a relatively simple geometry) but since it is necessary to determine the weights and centres of gravity of all such structural elements, the calculations for that purpose can be used to give values for the amount and location of the buoyancy contributions simply by taking the difference between the specific gravity of the material of the external structures and that of sea water.

In some instances the external components actually displace a greater volume than that of their material content. Thus, at the fore end, the structure may enclose tankage which is not necessarily flooded, while torpedo tubes are usually dry and sonar arrays are likely to displace volume in excess of their material content. Similar situations occur in the superstructure casing, bridge fin and stern. In the bridge fin, for example, there are the conning tower, periscopes and masts, though the snort induction and exhaust masts, which are buoyant when the submarine is snorting at periscope depth, are free flooding down to their hull valves when fully submerged. The propeller and shafting also have some buoyancy. The point is that all this miscellany of items have to be assessed, not just for their contribution to the buoyancy force (which is modest) but also to the centre of buoyancy position vertically and fore and aft (where their effect can be appreciable because of their leverage).

It is usual to assume in the buoyancy calculations that the MBTs are fully flooded and only their boundary structure displaces water. In practice, it is possible for air to be trapped in parts of the tank and contribute to the buoyancy. Air entrapment can arise due to the attitude of the submarine when diving: air could go to the forward or after corners of the tank depending on whether the boat initially sinks at a bow or stern angle, but it is usually taken that it dives with a bow down attitude. Since this air is highly compressible, with increase of depth, air pocketing can cause unpredictable and potentially dangerous variation in buoyancy. Thus in designing MBTs particular care has to be taken to provide adequate air venting arrangements, suitably located.

A similar problem can arise in free flood spaces at the fore and after ends and in the superstructure casing, where adequate provision for venting has to be made to ensure the air present when the submarine is on the surface escapes rapidly and completely during diving. However, this is not always compatible with the achievement of a low drag outer envelope because the provision of many holes and slots for quick venting – and also for quick flooding – of FF spaces can cause considerable added resistance, and so a compromise is necessary. The converse problem of ensuring that water freely escapes from these spaces during the operation of surfacing – which also encourages having plenty of vent and flood holes – causes similar conflicts.

The control surfaces and associated appendages are usually designed to be free flooding and provision made for the proper venting and draining of their enclosed internal spaces. An alternative approach is to fill these spaces with a buoyant material capable of withstanding full diving pressure, which not only overcomes any difficulty in venting and draining but also provides some protection to the inaccessible internal structure where corrosion can proceed unseen; moreover, the buoyancy provided by this means is a useful way of helping to counteract the weight of propeller, shafting and control surfaces at the extreme after end of the vessel.

Another buoyant volume which has to be taken into account is that of externally sited ballast, which may be stowed in a ballast keel, if fitted, or in MBTs, fore or aft FF spaces, or even in the superstructure casing. If stowed in MBTs, solid ballast will reduce the blowable volume and so affect the reserve of buoyancy.

3.12 So far we have focused attention on the displacement volume of the submarine. Buoyancy is, however, not only dependent on that volume but also on the density of the water it is displacing, and will vary if there is variation in sea water in the areas in which it is to operate. Sea water specific gravity (SG) can range from 1.00 near the mouths of rivers or in fjords where the water is virtually fresh due to melting glaciers, to about 1.03 where there is a high salinity. It is usual to work to a standard figure of 1.0275, which though pitched well towards the higher end of the density range is the average encountered in normal submarine operation. The design therefore has to cater for a variation of 3% in buoyancy due to sea water density variation, which represents quite a large change in the buoyancy force. Change may occur rapidly when approaching a coast or where there are substantial changes with depth (layers) of temperature and salinity, and the operating systems of the submarine have to be able to cope with them.

As well as these buoyancy changes due to sea water density variation, there can be buoyancy changes due to variation in the displacement volume of the submarine. An important example is the change in the geometry of the pressure hull caused by hydrostatic pressure loading with depth: as the hull experiences increasing pressure, its plating and frames experience predominantly compressive strain, resulting in a reduction in the diameter and length of the hull and of the displacement volume. The hull is fairly uniform in its compressibility, which consequently does not cause any significant shift in the centre of buoyancy; air pockets are highly compressible and so add to the effect – though if occurring forward or aft they could add to the problem by also tending to cause some centre of buoyancy shift. A similar compressibility effect can occur where the hull is treated externally with compliant tiles for acoustic reasons, as these too will compress as diving depth increases. The consequent buoyancy loss at full diving depth can be reckoned in tens of tonnes in a submarine of 2000 to 3000 tonnes submerged displacement.

3.13 This elastic compressibility causes an instability in depth. If it is assumed that at a certain depth a submarine is in exact neutral buoyancy and then increases its depth, the compression of the hull will reduce buoyancy, giving an excess of weight over buoyancy and a net force downwards tending to push the boat deeper and so cause a further reduction in buoyancy; if depth is reduced, the converse applies, and the effect is like that of a negative spring. To hold a prescribed depth requires positive control and some submarines are fitted with a hovering system for that purpose. The foregoing applies to moderate diving depths; if vessels are designed for great depths, which is likely to be for exploration, then it is possible for the compressibility of the water to result in an increase in density exceeding the compressibility of the hull.

The natural instability in depth due to compressibility is associated with uniform density sea water and could be reversed if there were a density gradient or layer in the sea characterised by the deeper water being more dense. If a suitable layer could be found, with a density increase with depth larger in magnitude than the compressibility effect, a submarine could acquire depth stability at the layer.

Other changes in buoyancy can occur intentionally in the operation of the submarine. An example is raising the periscope, which increases the displaced volume of the periscope tube external to the hull; it has been known for skilled submariners to hold a stable hover using a periscope, though this requires the submarine to be initially accurately 'in trim'.

Another effect which arises in boats fitted with external, sea-water compensated, diesel fuel tanks may be regarded either as a change of buoyancy or of weight. These fuel tanks operate at ambient sea pressure and when full, displace sea-water volume with lower density diesel fuel. As fuel is consumed it is replaced by denser sea water and the change can be treated as an increase of weight or a reduction in buoyancy due to the reduction in fuel displacement. This change also has been utilised to afford a means of hovering by transferring fuel between internal and external tanks.

When a submarine moves from sea water of one density to another, the entrained water in MBTs and FF spaces is unlikely to change much, and in consequence calculation of the effect of changes of sea-water density on buoyancy is related to the outer envelope of the submarine rather than to the net displacement of the pressure hull plus miscellaneous external volumes.

WEIGHT ELEMENTS

3.14   In the next chapter we consider in some detail how weight and space considerations influence submarine design – they have a very important part to play – and for now our concern is more limited: to indicate the relationship between weight and buoyancy in the present context of hydrostatic stability and flotation. As we will show, that too is an important aspect of submarine design. Weight is, of course, a vital parameter in the design of all vehicles, particularly those which depend on dynamic lift to support them such as aeroplanes and hydrofoil craft.

Although, as we have already observed, buoyancy is a more reliable source of support, the need to keep buoyancy and weight in or close to balance in submarines is demanding on the designer.

The determination of the weight of a submarine is more difficult than that of buoyancy. Whereas buoyancy is primarily related to gross volumes of geometrically simple forms, weight is the summation of every element that goes into a submarine. Large components of the fixed weight are the pressure hull structure, main propulsion plant, batteries and stowed weapons, and these can be evaluated relatively easily. However, they are not as dominant as the pressure hull volume is in relation to buoyancy. The difficulty lies more in all the minor structure, systems and equipment to be installed for which reliable weights are hard to obtain. In addition to the permanently installed components there are the variable weights associated with crew, their effects, stores and the fluid contents of many tanks.

For the present, our interest in weight is for its gross properties in a submarine, namely, the all-up weight and centre of gravity positions longitudinally and vertically; and also in how those properties vary in normal circumstances in the operation of the submarine. Before dealing with how variation occurs and is corrected, it is pertinent to examine the reason for having ballast – whether permanent or variable – in a submarine.

3.15 Starting with permanent ballast, it might seem inefficient for a submarine to carry round such deadweight. However, permanent ballast can be necessary in design for one or more of the following reasons:

(a) To assist in bringing about equality of weight and buoyancy in the submerged condition. (It is unlikely that the totality of necessary weights in the submarine will naturally equate to the summation of the buoyant components, and in design it is prudent to provide some excess of buoyancy to allow for eventual adjustment of weight, the other way round being impractical).

(b) To enable the longitudinal positions of centre of gravity and centre of buoyancy to be brought into coincidence. (It is unlikely that the longitudinal distribution of weights in the submarine when its contents are best arranged for efficient working will accord with the longitudinal distribution of the buoyancy components). (Figure 3.7)

(c) To bring the vertical position of the centre of gravity sufficiently below that of the centre of buoyancy to give an adequate separation for the purposes of transverse and longitudinal hydrostatic stability. (A corresponding comment to that on longitudinal distributions of weight and buoyancy applies).

The first of these reasons brings to attention the issue of design margins which we take up later. For now we would just observe that this reason is usually the least telling in determining the need for permanent ballast. The second reason is more significant because there is a disparity in modern 'streamlined' submarine forms between the trend for the centre of buoyancy to be forward of mid-length (due to the full bow and fine stern) and the centre of gravity to be aft (due to heavy propulsion machinery aft and relatively light spaces forward for control room, electronic spaces and accommodation). In diesel electric submarines, the heavy weight of the batteries – which can be located forward – con-

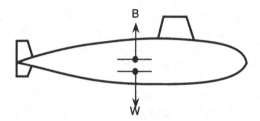

(a) In longitudinal balance   LCG ≡ LCB

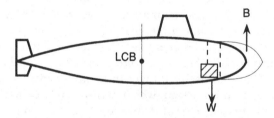

(b) Additional weight forward requires added
     buoyancy forward

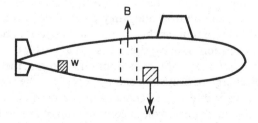

(c) Added buoyancy not located at added W т.W
     additional Wт.w necessary and B = W + w

Fig. 3.7   Maintenance of longitudinal balance

tributes to reducing the longitudinal balancing moment needed from permanent ballast, but in submarines with nuclear plant a substantial moment can be called for which might necessitate several hundred tonnes of permanent ballast for balancing purposes alone.

As regards the third reason, the geometry of a cylindrical hull causes the centre of buoyancy to lie close to the axis of the cylinder and, by the same token, the distribution of weights in the submarine (machinery centred about the axial shaft usually adopted, and spaces forward of more or less uniform density) leads to the uncorrected centre of gravity also lying close to the axis. The need then is for permanent ballast as low down as can be achieved to pull the overall centre of gravity below the axis. In this regard also, the batteries in a diesel electric submarine help as they can be located in the lower part of the hull; in consequence, it might be possible in the design of submarines of that type to dispense with permanent ballast altogether, though that would call for much skill and experience on the part of the designer.

3.16   In the interests of consistency, we touch briefly here on the second type of ballast, namely, margin ballast, though it is treated at greater length in the next chapter in the discussion on design and other margins of weight and space. Margin ballast is about making an allowance for weight growth in a submarine during design and building and for its future life in service. Whereas permanent ballast really does have to be stowed in the completed submarine, margin ballast might or might not depending on whether the designer has been successful in his budgeting for weight control and if, therefore, there is something left in hand for future growth. What is advisable is that a prudent designer should aim to cater for uncertainty in his design, which in turn is related to the degree of repetition of past practice or of innovation, i.e. to the evolution/revolution mix to be incorporated in his design.

The third type of ballast, variable ballast, is unlike the other two – which are stowed in the submarine in solid form, usually lead blocks – in being carried as sea water. This is because it is used to make continuous corrections to achieve the neutral buoyancy condition, as regards both force and longitudinal moment, as the weight and buoyancy change with the submarine in operation. During the course of a patrol, the weight of the submarine will vary due to the consumption of stores, fuel, distilled water, etc., and discharge of brine and sewage, etc. Also the buoyancy of the submarine will vary due to the changes in the density of seawater between different locations on patrol, to compression of the pressure hull at depth, and (if applicable) to externally stowed diesel fuel being replaced by sea water. The means usually provided for adjusting the state of the submarine to allow for all these effects is to make changes to the weight of the submarine by admitting or discharging sea water from tanks inside the pressure hull. The tanks constitute the main part of the Trim and Compensation System and their contents constitute the main part of the variable ballast. Because they play a central role in the process of maintaining a submarine 'in trim', a term with special meaning to submariners as to aircrew, we now go on to give the tanks particular attention. We should point out that this special usage of the word 'trim' is confusing because generally in a marine context it is used to mean the slope of the keel of a vessel, represented in surface ships by the difference in draughts forward and aft. The only advice we can give is: judge from the context which meaning applies.

TRIM AND COMPENSATING TANKS

3.17   To submariners, being 'in trim' means that the submarine is as nearly as can be contrived in neutral buoyancy and also has its longitudinal centres of gravity and buoyancy in the same vertical with the axis of the submarine horizontal. Submarines in the past, when obliged to submerge, often had to proceed slowly for quietness and to conserve battery power. To be accurately 'in trim' was a vital necessity, and even for the modern submarine there is always advantage in keeping well in trim. Because of its significance from that aspect, and because it also gives a glimpse of how hydrostatics and hydrodynamics interact in submarine operations, which is particularly relevant to the submarine designer, we devote Appendix 2 to Operational Practice for Catching and Keeping in Trim.

Of direct concern to the designer is how to arrange the Trim and Compensation Tanks to best advantage. A basic method is to have just two tanks, one well forward and one well aft in the hull, usually termed the forward and after trim tanks (Figure 3.8(a)). To correct for a change which only affects longitudinal balance, one tank would be partially emptied and the other partially filled, thus keeping weight constant while changing the longitudinal moment; to correct for a change which only affects weight, one or both of the tanks would be partially emptied or filled while keeping the longitudinal moment constant; to correct for changes which affect both weight or buoyancy and longitudinal moment, other combinations of flooding/emptying the two tanks would be appropriate.

Because the tanks are usually located within the pressure hull, they can be readily enough flooded from sea, with the tanks venting to submarine atmosphere. Discharging water from the tanks is not so easy; water can be blown out to sea using high pressure air, but that is noisy and subsequent venting of the tanks raises atmospheric pressure inside the boat. It also requires the tank structure to be "hard" i.e. capable of withstanding diving pressure. Alternatively, the water changes can be made using a pumping system to discharge to sea, which either requires a dedicated pump at each end of the submarine or a centralised pump with inlet and discharge valves located together with pipes running to the forward and after trim tanks, all the piping being required to withstand diving pressure.

The three tank system is a variant of the foregoing in having a central tank amidships – the Compensation Tank – as well as the Forward and After Trim Tanks (Figure 3.7(b)). This arrangement makes it possible to use only the central tank for discharge to the sea or taking in water from sea, and for the Forward and After Trim Tanks to be entirely isolated from the sea, using them only for adjustments to longitudinal balance. To maintain the isolation of the trim tanks a two pump system will be used, one a large overboard discharge pump connected only to the compensation tank and the other a small trimming pump to transfer water between the trim tanks. In this way the amount of piping at diving pres-

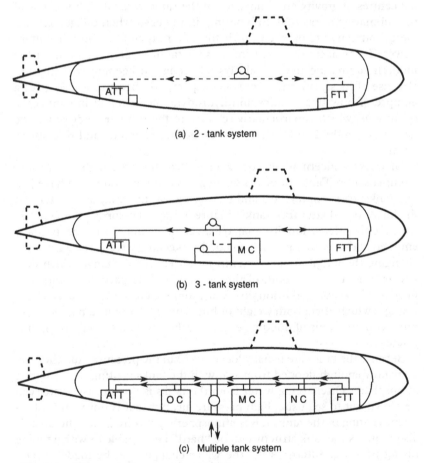

(a) 2 - tank system

(b) 3 - tank system

(c) Multiple tank system

Fig. 3.8   Trim and compensation systems

sure inside the submarine can be much reduced, while only the compensation tank and its overboard piping need to be 'hard'.

This desirable three tank arrangement might not be practicable if there were insufficient space in the lower part of a submarine for tankage in the appropriate locations. In those circumstances, it would be necessary replace the single compensation tank near amidships by several smaller tanks located wherever they could be conveniently fitted (Figure 3.8(c)). To avoid such an arrangement leading to a more complex piping system exposed to diving pressure, one of these smaller compensation tanks could be made the 'hard' tank and the others made 'soft' tanks, the latter interconnected to the trim tanks by 'soft' internal piping systems.

SPECIAL TANKS

3.18 As well as Trim and Compensating Tanks, submarines are also provided with tanks for special purposes. The contents of these tanks may or may not be treated as variable ballast, as noted below.

Weapon Compensation Tanks

For a submarine armed with torpedoes there is a significant change in weight when they are discharged. It would be highly undesirable if, during the firing sequence at or near periscope depth, the balance of the boat were disturbed as this might lead to the bow breaking surface or at least a loss of accurate control of depth. To ensure that this cannot happen, several compensation tanks dedicated to weapon discharge are provided. One is the Water Round Torpedo (WRT) Tank, which contains an amount of water sufficient for a full salvo of torpedoes. When a torpedo is loaded into a torpedo tube, and the rear door closed, and prior to opening the bow cap, water from the WRT tank is admitted to fill the space between the torpedo and the tube, which can then be equalised with sea pressure and so allow the bow cap to be opened with no change in weight or longitudinal moment as the WRT water has only been moved vertically. Once the torpedo has been discharged, the tube will fill with water and so replace the lost volume of the torpedo. If, as is usual, the torpedo is denser than water there is a need to take in more water to replace the lost weight and for this purpose a second tank, the Automatic Inboard Venting (AIV) Tank, is provided into which an appropriate additional amount of water is admitted. The submarine is thereby able to keep in balance, though now with a flooded tube. To reload another torpedo it is necessary to shut the bow cap and drain down the water from the tube into a third tank suitably located to continue to maintain longitudinal balance; this is the Torpedo Operating Tank (TOT) and its capacity is sufficient to take all the water required to compensate for the loss of weight which would occur if a full load of torpedoes were discharged. The means are also provided for readily transferring water from the AIV Tank to the TOT and from the TOT to the WRT Tank. The contents of these tanks are not treated as variable ballast.

D Tanks

These are tanks located near the longitudinal centre of buoyancy of the submerged submarine and contain water so that they are available to compensate rapidly for loss of buoyancy due to compression if the submarine were to dive deep quickly. D Tanks are not always provided, in which case reliance would have to be placed on Trim and Compensation Tanks but, if so, there could be difficulty in keeping in trim if the submarine were to proceed very slowly having suddenly dived deep, because of the slow rate at which water could be pumped out at depth. If D Tanks are provided, they are made 'hard' and fitted with a high pressure air blow to enable them to be quickly emptied at depth. As they are essentially compensation tanks, D Tanks are sometimes incorporated within the boundary of the main compensation tank, in which case they can serve as the 'hard' tank element of that tank. Because compressibility effects work both ways the D Tanks are two tanks arranged so that one is full and the other empty or both half full in normal diving conditions. The contents of D Tanks are treated as part of the variable ballast.

Q Tank

This tank was an essential feature of older diesel electric submarines which were obliged to spend time operating on the surface, its purpose being to aid rapid submergence in a crash dive. It is still found in some submarine designs today as it is a powerful way of getting away from the surface quickly. Q Tank is a relatively small tank located well forward in the hull, which is normally empty and when flooded brings the bow under the water so that the submarine can quickly drive itself down. It is effective for the purpose both because of the bow down moment and of the negative buoyancy it causes, but that does mean that as well as being provided with a large hull valve for rapid flooding it also requires a high pressure air blow to enable it to be rapidly emptied again as soon as the submarine has gained adequate depth. It is not normally treated as part of the variable ballast and is assumed empty for design purposes. In some designs it is known as a safety tank and will normally be kept flooded whilst submerged. In this mode rapid discharge aids recovery from control or flooding accidents.

Hover Tanks

As we saw earlier in the chapter, a submarine is vertically unstable in depth when stationary because of compressibility effects. If, operationally, there is a requirement for the submarine to keep a set depth whilst stationary, i.e. to hover, then some form of control has to be introduced into the buoyancy/weight equation, which can be achieved by flooding in water as the submarine starts to rise and pumping water out as it starts to sink. To achieve control requires a means of rapidly moving water in and out of the boat to counter its natural motion, and this can be effected by providing special tanks dedicated to this purpose of hovering. These may be pressurised to ambient sea pressure so that transfer of water can be readily accomplished against zero differential pressure head. Only relatively small amounts of water are involved in this process when hovering when the submarine is at depth, and use has been made of the interchange between internal and external fuel tanks, when fitted, so that oil fuel flow alters the water displacement of the external fuel tanks. If, however, the requirement is to hover near surface, then the action of waves causes an additional destabilising effect and the hovering system has to be more powerful to cater for the suction forces generated as the waves pass over the hull. We discuss the control requirements of the system in the later chapter on submarine dynamics.

HYDROSTATIC STABILITY

3.19 To round off this chapter, in which we have described at some length the wealth of hydrostatic aspects which bear on the design and operation of submarines, we turn to consideration of their stability – not in a dynamic sense, which is also relevant and which we treat later, but in a static sense. Recognisably, and not just for submarines, the separation of static and dynamic considerations is rather artificial, for in reality the two aspects often occur concurrently, but it is a convenient way for dealing with matters which, while fundamental, are by no means simple.

In a submerged vessel, hydrostatic stability, both transverse and longitudinal, i.e. in heel and pitch, requires that the centre of gravity should be below the centre of buoyancy. It is therefore the size of this distance, BG, which governs the stability characteristics of the submarine.

There are consequences for the submarine, particularly in design, of providing hydrostatic stability both when submerged and when on the surface.

Submerged stability

For the submerged submarine we need only consider intact stability since – because there is no reserve of buoyancy – any damage leading to flooding will cause the submarine to sink unless emergency recovery action could bring the damaged submarine to the surface. As we have already seen, submerged stability just requires that the centre of gravity G should be below the centre of buoyancy B, (Figure 3.9) and so the significant question in design is, by how much? For the magnitude of BG determines the restoring moment when the submarine experiences an angle of heel or an angle of pitch. We now go on to discuss criteria for the magnitude of BG in relation to several circumstances which put it to the test.

(a) **Static:** it is desirable to have sufficient BG that moving weights or men on board athwartships or fore and aft does not cause a large change of attitude.

(b) **Dynamic:** the value of BG affects the behaviour of the submarine when proceeding at speed submerged. If it changes course at speed there is a tendency for the bridge fin, which acts as a lifting surface at the top of the hull, to cause a heeling moment; depending on the size of BG, the result could be a very large heel angle, sometimes termed 'snap roll' as it occurs rapidly shortly after the rudder is put over. Although the bridge fin can be designed to reduce the lifting force it experiences during a turn, the contribution of the hydrostatic moment associated with BG to resisting heeling is also important and sets a lower limit on BG.

By the same token, BG provides a hydrostatic moment to the hull resisting pitching. As we discuss in detail in the chapter on Submarine Dynamics, BG contributes what is

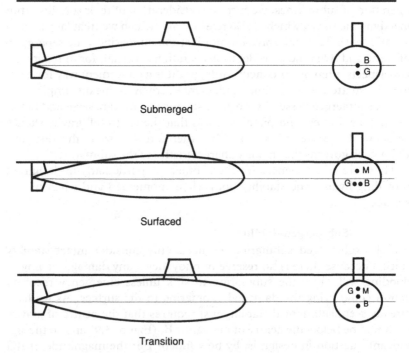

Fig. 3.9   Cross-section of submarine and changes in B, G and M between surface and submerged

effectively a spring stiffness term to the equations of motion in the vertical plane, leading to oscillatory characteristics in the pitching motion. Because that contribution is hydrostatic and all the other forces and moments are hydrodynamic in character, its effects on the motion change with speed, resulting in different control characteristics of the submarine at different speeds. At very slow speeds, the dominance of the hydrostatic moment results in what is termed the 'Chinese effect' of the depth control surfaces whereby the after hydroplanes no longer cause the boat to dive when put to dive but instead cause it to angle down by the bow while experiencing an upward rising motion. Whether this effect is significant depends on the magnitude of BG.

(c) **Transition:** another aspect influencing decision on the magnitude of BG is the transition from the submerged to the surfaced condition, as the transverse metacentre moves from the centre of buoyancy to the position appropriate to the surface waterline when it is fully established. While this is happening, the centre of gravity also moves owing to removal of water from the MBTs, which gives a free surface effect. The free flood spaces are also important during transition to surface, for if the submarine surfaces rapidly they cause a delay in draining down water and there is an effective raising of the vertical centre of gravity submerged which, though temporary, may lead to problems. It is necessary to ensure that during transition the migrating centre of gravity is not above the metacentre, which also changes.

A particular variant on the theme of transitional stability is associated with the circumstances encountered when a submarine surfaces through ice. Even if the under surface of the ice is reasonably flat, the force it will apply at the top of the bridge fin before break-through occurs has an adverse effect on transverse stability. The force can be limited to some extent by controlling the rate of ascent of the submarine, but there is sufficient reason for concern to take account of the operation in design when making judgements about the magnitude of BG.

Calculations for the vertical and longitudinal positions of the centre of gravity are usually based on the identified weights and their location. Strictly, when the hull moves, everything within the outer envelope moves including the water contained within free flood spaces, casing and bridge fin. However, as the latter are exactly compensated by their buoyancy these components of weight are usually ignored for static purposes, though they have to be included in the mass and inertia of the vessel for dynamic purposes. They should be taken into account when the submarine moves into other density water, but it is the practice to conduct stability analysis in standard density conditions.

In design it is usual to find that application of the foregoing conditions does not give definite guidance on what BG should be. The fact that in this field 'the bigger the better' generally obtains also does not help. Rather the designer would be more helped by knowing how little he could safely get by with, since that would avoid an unduly high price in amount of permanent ballast to be stowed. In the circumstances, as would be expected, the designer is likely to revert to consistency with previous practice. The outcome will depend on the type and size of submarine he is designing, and a proportion for BG in relation to the diameter of the pressure hull of 3 to 4% is not untypical.

Surfaced Stability

3.20   It is not to be expected that much emphasis will be placed on the stability of the modern submarine on the surface because it is unlikely to surface operationally except when entering and leaving coastal waters at the beginning and end of a patrol, though it may very occasionally be on the surface when on patrol for particular and unusual purposes. Generally, modern submarines can be regarded as *Albacore* derivatives and as such they have poor surface performance in all respects, including transverse stability and, come to that, longitudinal stability too, since the waterplane on the surface is insufficient to cause the longitudinal metacentre to be much higher than the transverse metacentre.

The position of the transverse metacentre will depend on the geometry of the outer envelope: if the MBTs are contained within an outer envelope which is of circular form because of the solid of revolution geometry mainly used, the metacentre will be on the axis – as is the centre of buoyancy submerged – and then the difference in metacentre height between the submerged and surfaced conditions will depend on what happens to the centre of gravity. Although, as we discussed above, the centre of gravity might effectively rise in the transition from submerged to surface condition, for geometries of the axisymmetric type it usually settles down on the surface to a position not far from that submerged, in which case GM and BG will be about the same. Many other geometries are, of course, possible with wrapround, underslung and saddle MBTs; most are likely to have surface condition metacentres rather above the axis, but not to any great extent.

As regards hydrostatic stability at large angles, the curve of righting levers to base angle of heel, θ, i.e. the GZ curve, of the submerged sub-marine is close to a sine curve, so that $GZ=BG \sin\theta$, which represents a good range with maximum GZ at 90° heel. For the surfaced submarine, the character of the GZ curve will depend on its external geometry, that is, whether it is axisymmetric or not, and on how the upper boundaries of the MBTs are shaped. Generalisation would, therefore, be difficult, although it can be readily recognised that because submarines on the surface have very small reserves of buoyancy compared with surface ships, the GZ curve is likely to be characterised both by small maximum value and range. (On the other hand, with its small profile, the surfaced submarine does not face as much in the way of rolling excitation from wind and waves as most surface ships).

REVIEW

From the foregoing discussions it will be seen that the submarine designer has to cater for two major flotation conditions, surfaced and submerged. To bring about the change, part of the total volume of the vessel must be contrived as floodable tankage, the Main Ballast Tanks. These must be sized to provide adequate additional buoyancy to give a satisfactory surfaced condition. This includes a longitudinal distribution of the MBTs so that the submarine floats nearly level and is of a shape which provides adequate transverse stability on the surface.

Submerged, the MBTs are totally flooded and the main contribution to buoyancy is the pressure hull volume. To achieve the neutral buoy-ancy condition requires an exact balance of weight and buoyancy forces which brings about a direct link between weight and volume of the wholly submerged vessel. Once the design is committed the buoyancy volume of the hull is fixed and there is no scope for additional weight which in a surface ship can be accommodated by additional sinkage of the hull, i.e. increase in draught.

As both weight and buoyancy can vary during the course of a patrol, it is necessary to provide special tankage within the hull, the Trim and Compensating Tanks, to adjust the vessel to neutral buoyancy.

In addition to the two primary conditions some account must be taken of the transition flotation states of the vessel as it submerges or surfaces. In particular, the transverse stability of the vessel may become marginal in such transition states and so special care is necessary in assessment of adequacy of stability.

4 THE WEIGHT/SPACE RELATIONSHIP

PURPOSE

4.1 We start this relatively brief chapter with an explanation of its purpose, because it is different in character from the other technical considerations involved in submarine design. In some ways it is not especially technical at all, but rather akin in nature to the debates on spatial design which architects indulge in. The issues which arise in consideration of the weight/space relationship for submarines might appear at first sight to be simple – they certainly are very basic – but that is deceptive because they become progressively more complicated as the relationship is explored in greater detail. Although the subject of weight and space and how they are related in submarine design is associated with hydrostatics, it goes beyond what can properly be treated under that heading because of the somewhat intangible nature of the relationship and its consequences in some regards as compared with the more matter of fact nature of hydrostatics.

The chapter is ultimately about 'what determines the size of a submarine'? In a particular submarine design, does it have to be of a certain size to provide enough buoyancy to support its weight or does it have to be of that size to provide enough space for its contents, so that it then has more than enough buoyancy to support its weight? If the former, there would be some space to spare, so how could the extra space be utilised? If the latter, weight in one form or another would have to be added to enable the submarine to attain the same density as sea water when it submerged and, would that be inefficient?

Our aim is to show how those questions may be answered. Central to the aim is the property of the submerged submarine that it is the densest of all marine vehicles and so next we consider the role of density in the design of marine vehicles that float.

SIGNIFICANCE OF DENSITY

4.2 Density in the present context is significant as it is the parameter connecting mass and volume, which in terms of weight and space are the relevant measures of the elements of which a submarine is made up. As described in the previous chapter on hydrostatics, a vessel which floats on the surface of the water has a reserve of buoyancy because of the free-board of the main hull and of any superstructure which is watertight. In consequence, the total volume of the vessel is larger than the volume underwater, i.e. its volume of displacement. It can be seen that if the reserve of buoyancy is R, as a ratio of displacement, then the ratio of the volume of displacement to the total volume is $1/1+R$. Because the buoyancy due to the volume of displacement equals the weight of the vessel, that ratio is also its specific gravity relative to sea water, which provides a measure of the overall density of the vessel, regarded as its weight divided by its total volume below and above water. (Figure 4.1)

Most surface warships have a relatively low overall density; thus for a frigate the ratio $1/1+R$ is typically around 0.3 (some 70% of the total volume being above the waterline) and for an aircraft carrier the ratio is around 0.2 (about 80% of the total volume being above the waterline). Passenger liners and container ships are also low density vessels; they are not dissimilar to the warships as regards high proportion of total volume above the waterline. Loaded bulk carriers and oil tankers on the other hand are relatively dense vessels, their $1/1+R$ ratio being around 0.8 (only about 20% of their total volume being above the waterline). As we have commented elsewhere, these characteristic proportions are related to the density of the cargo carried by a vessel: the denser the cargo, the higher will be the overall density of the vessel. That might seem trite, but the extent of the difference can be surprising.

Now a submarine on the surface typically has, as we have seen, a reserve of buoyancy of only 10%, so that the ratio $1/1+R$ is 0.9, which is already very dense, while for the submerged submarine the ratio is 1.0, which is as dense as can be achieved with neutral buoyancy. Several significant consequences stem from this fact. One is the considerable difference in space available in the submarine compared with a surface war-

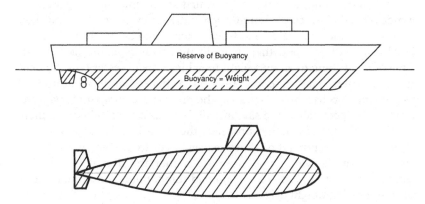

Fig. 4.1   Volume and displacement

ship of the same displacement; the frigate we referred to earlier has 3.5 times the total volume of the comparable submarine, while a ballistic missile submarine of 15 000 tonnes submerged displacement will have no more space than a frigate of around 4 000 displacement.

4.3   Another consequence is less obvious: the high overall density of the submarine is not due to its cargo being of exceptionally high density; on the contrary most compartments in a submarine as we show later have densities not dissimilar to those found in surface warships – though battery compartments in diesel electric submarines and reactor compartments in nuclear submarines do approach sea water in average density. There is not enough permanent ballast to account for the submarine's high density, and the explanation lies in the pressure hull, and the fact that in <u>most</u> cases a submarine needs to have a heavy pressure hull in order to reach an overall density equal to that of sea water. We pursue that aspect further in the chapter on submarine structures and just for now restrict comment to the observation that in submarines there is a choice which does not exist in surface ships. The loading due to waves which primarily governs the requirements for strength and stiffness of a surface ship's hull structure is determined by Nature and the designer must conform to its demands. For submarines, the primary source of loading is hydrostatic pressure at deep diving depth. It is available hypothetically to the designer – in advising on the trade-offs influencing decisions on operational requirements for diving depth capability – to vary the pressure hull structure at will. The questions that then arise are to what extent and in what ways does manipulation of pressure hull structural weight (and with it diving depth capability) affect the design of the submarine as a whole? It must be recognised that, while diving depth might be a trade-off at Concept Design, once fixed, hull weight/diving depth can no longer be used to achieve weight balance.

The authors contend that to a good first approximation the generality of modern submarines are initially space driven in design, i.e. that their size is determined by the demands for space which derive from the functional requirements, and that their diving depth capability is the outcome of fine tuning to all up weight to achieve the necessary overall density of sea water. The contention rests on the belief that the magnitude of deep diving depths which emerge in the process are compatible with the operators' expectations. We say 'initially' in this assertion because there is little doubt that as design proceeds the discipline of keeping within weight budgets in pursuit of effective weight control (which we return to later) will give a widespread impression that the design is – or has become – weight limited. That is no bad thing, but it is not the same as the design being weight driven.

It is of course readily conceivable that instances could occur in which a submarine design would be weight driven, if there were a requirement for an exceptionally large deep diving depth. In a manner of speaking such a design would lack harmony between its all up weight and total space – certainly the volume required to support the exceptionally large weight (due to the heavy pressure hull structure) would by definition be larger than necessary to meet the space demands. We suspect that what could then happen would be a developing belief that the design was inefficient, leading to a search for ways in which to put the extra space to good use. The contents would add to the weight, necessitating some reduction in the exceptional diving depth and the outcome, we suggest, would be a submarine which had regained harmony as regards its all up weight and total space but which, had it started with extra contents in from the beginning, would have appeared to be space driven.

To avoid this overlarge design solution an alternative design strategy could be adopted. This is to assume, as advocated, that the design is space driven. This would establish a hull size compatible with the space/volume requirement. The subsequent assessment, including the large diving depth requirement, would reveal an excess of weight over buoyancy. A discussion could then ensue as to how balance could be achieved. A relaxation in depth requirement might be sought pointing out the cost implications; higher strength/weight materials may be considered for the hull; reduction in speed/endurance might enable a trade-off between propulsion and structural weight budgets; or, more likely, a combination of such solutions plus an increase in size may lead to a satisfactory solution.

We now point out that in this discussion of size determinants in submarine design we have – deliberately – oversimplified the issues. There are others which require attention like the impact of a single versus double hull configuration, i.e. wrap-round external hull; how choice of pressure hull structural material affects the outcome; the balance of choice between keeping as much equipment as possible inside the pressure hull and locating some of it outside; and the benefits or otherwise of a strictly austere policy as regards space provision (the 'shoe horning' approach). Our justification for oversimplifying at first is that the characteristics of being space driven or weight driven represent the ends of a spectrum within which will lie all possible submarine designs. What determines where a particular submarine design will lie depends largely on the other aspects just listed, and these we return to after going into more detail about basic weight and space matters.

WEIGHT

4.4 We go on to consider the proportions of weight (and space) devoted to the various functions of submarines. Because of the dearth of published information on this and other data for submarines, it is neces-

Table 4.1 *SSK*

	Weight (%)	Space (%)
Payload	9	28
Structures	43	—
Main and Auxiliary Machinery	35	56
Accommodation and Outfit	4	11
Stores	1	5
Permanent Ballast	8	—

Table 4.2 *SSN*

	Weight (%)	Space (%)
Payload	8	30
Structures	45	—
Main and Auxiliary Machinery	35	55
Accommodation and Outfit	4	10
Stores	1	5
Permanent Ballast	7	—

sary to be somewhat creative in compiling useful comparative lists, but we believe the following tables, Table 4.1 for a modern diesel electric, SSK, submarine and Table 4.2 for a nuclear attack, SSN, submarine are reasonably representative. The tables compare the percentages of 'dry' weights related to standard displacement (that on the surface without variable loads) and of volumes of 'dry' spaces. Both types of submarine are *Albacore* derivatives and so are of axisymmetric form and single hull configuration.

What can immediately be recognised is that, despite the considerable differences in main and auxiliary machinery between diesel electric drive and nuclear propulsion plant, there is remarkably little difference in the weight and space proportions for the SSK and the SSN. Thus in what follows we can dispense with differentiation between them. We should mention that for present purposes we have taken the payload of these military submarines to be weapons stowage and launching arrangements together with control room, operations spaces, sensors and their offices.

It will be seen from the tables that about half the total weight of the submarines in standard condition is taken up by structure; although not indicated, well over half of that component will be pressure hull structure – and since it takes up very little space it is a relatively dense item. In the submarines represented by these tables almost all structure would be of steel, though in the SSK some external structure like the superstructure casing and bridge fin sheathing might be in glass reinforced plastic for top weight reduction reasons.

The main and auxiliary machinery contribute around a third to the total weight in the standard condition. In the SSN it can be inferred that the reactor compartment with its massive lead shielding – provided to protect personnel afloat (and ashore) from radiation hazards – represents about half of the machinery component. In the SSK, the batteries occupy a proportion of the machinery weight similar to that of the reactor compartment in the SSN, though there is greater freedom in vertical disposition of batteries than of the reactor because of its size.

The permanent ballast, which like structure takes up little space, is another very dense item for which there is freedom in vertical disposition, exploited in design to pull the vertical centre of gravity of the submarine as a whole down below the axis for hydrostatic stability purposes.

The payload weight item might seem small in view of the fact that it is what the submarine exists to deploy, but the smallness is due to the relatively low density of electronic compartments and torpedo stowage spaces.

4.5 Although we are primarily concerned in this section with weight aspects, it is worthwhile to continue for a while with also looking at the space proportions in the tables because of the light they throw on the issue of the densities of the components of the submarines. It will be seen from the tables that the space given over to the payload, at around 30% is a reasonable proportion of the total. Moreover, if one links that figure with the 10% or so occupied by accommodation and outfit – which is a measure of the space devoted to the crew – on the grounds that together they represent most of the 'fight' part of the traditional set of warship purposes (to float, to move, to fight), then the 40% sub-total is arguably a respectable achievement. In passing, while on the subject of submarine accommodation, the 10% of total space is far smaller than that found in surface warships – where 30% is not untypical which, it should be remembered, is related to a much larger total space. The difference is mainly due to the fact that submarines generally operate with much smaller crews than surface warships; thus, for example, the complement of a 2500 tonnes SSK will be around 50 officers and ratings, whereas the corresponding number for a frigate of that displacement is around 150. Apart from payload plus accommodation, the other dominant space item is the volume given over to the main and auxiliary machinery, at over half the total 'dry' volume. This is a higher proportion than is found in surface warships – where again 30% is not untypical, though on a higher total volume – and reflects the submarine's need to carry stored energy (whether in a reactor core or in batteries) and to provide means of controlling the atmosphere, offset by absence of uptakes and downtakes. Nevertheless, it is striking that over half a submarine's volume and about a third of its weight, excluding liquids, are taken up by main and auxiliary machinery.

4.6 Tables 4.1 and 4.2 can be used as guides to the densities of the contents of a submarine, by considering for each item the ratio of its weight percentage to its space percentage – if this was unity, the item would be as dense as sea water, while the lower the ratio the less dense the item will be. Leaving aside structure and permanent ballast which are palpably dense, the next in order is the main and auxiliary machinery for which the ratio is around 60%. Now that is an exceptionally high figure, which is due to the relatively very dense reactor compartments in the SSN and battery compartments in the SSK, which as we have previously observed are not much less dense than sea water. If one excluded those spaces, the ratio for the remaining main and auxiliary machinery spaces would drop to around 40%, which is not very different from the ratios for the other items in the tables. Significantly, excluding structure and permanent ballast, overall the ratio for both SSK and SSN is around 50%, so that the contents of the submarines average out at about half the density of sea water. Hence our hypothesis of the role of the pressure hull – as the main structural component – in bringing about the property of the submarine that it is the densest of marine vehicles.

WEIGHT ASSESSMENT AND CONTROL

4.7 After that digression into density, in which we looked at the weight components of a submarine in very gross terms, we now examine how weight is treated in design in the detail necessary for the achievement of confidence that its assessment will be realistic and sufficiently accurate for the purposes of the design phases. In this respect submarine design is not different in kind from surface warship design, though it is different in degree for the reasons we have already commented on. If the eventual all-up weight of a newly designed and built surface ship exceeds that allowed for in the design calculations, including any margins, the ship will float at deeper draughts. There will be, possibly, some loss of stability if the excess weight occurred above the design centre of gravity – there will also be penalties in performance as well. Unless all-up weight was very seriously undercalled or its centre of gravity badly misjudged the outcome would not be intolerable. In a submarine submerged, however, there is no reserve of buoyancy to eat into in a corresponding situation and the only way out would be to reduce the amount of permanent bal-

Table 4.3 *Weight groups for a submarine*

Group 1	Hull structure
Group 2	Propulsion systems
Group 3	Electrical systems
Group 4	Control and communications
Group 5	Ship services
Group 6	Outfit and furnishings
Group 7	Armaments and pyrotechnics
Group 8	Fixed ballast
Group 9	Variable items

last with an inescapable reduction in hydrostatic stability wherever the excess weight occurred. In the latter event the limits of tolerability could easily be exceeded and then the only available course of action would be to increase buoyancy. This could be done only as a paper operation by lengthening the submarine, provided the need was revealed sufficiently early on in the design process. There would be a consequent upheaval to the internal arrangements of compartments and an impact on many other aspects of design including structure, manoeuvring and control, propulsion, etc.

The criticality in submarine design of achieving accurate weight assessment is, it will be recognised, a feature which it shares with vehicles which depend on dynamic lift to support their weight, and calls for the same disciplined and meticulous attention to detail. The fact that buoyancy is a more dependable form of support than dynamic lift is more than offset by the use of prototypes to confirm the solution in the design and production of vehicles which utilise dynamic lift. The essential emphasis in submarine design has therefore to be on providing guaranteed safeguards against undercalling in weight assessment, and we now go on to discuss what those safeguards are and how they can be achieved.

4.8   A basic safeguard is what one might call sound bookkeeping, i.e. to contrive a way in which all weights in the submarine can be fully accounted. For many years now submarine designers have used a weight grouping method intended to be sufficiently comprehensive and taken to a level of detail such that every component of a submarine can eventually be located within it. The method, originally developed by the US Navy, uses a four or five digit level categorisation, the ultimate level depending on how far it is necessary to go to identify the individual parts of a particular system or sub-system. The first level groups are shown in Table 4.3 from which it can be seen that although they have an apparent logic there is nevertheless some arbitrariness in them – this is, in part, explained by the historical development of the groupings and adaptation to new situations. However, the main reason is that there is such a close and complex inter-relation between elements that any subdivision inevitably leads to anomalies.

Another measure is investment in a weight data base of high quality, which goes hand in hand with the weight grouping method, though still requiring experience and understanding on the part of the custodians of the data base so as to ensure that cause and effect are properly attributed. The worth of the data base would be heightened if a policy of progress in submarine design by evolutionary development could be pursued, since clearly risk in weight assessment increases in proportion to the extent of innovation and departure from previous practice, but a policy of that nature is rarely for the designer to command.

Another safeguard is achieving conviction on the part of all those who contribute to creating weight – including weapons, machinery, equipment,system designers and the submarine builders and their sub-contractors – that accurate prediction and meticulous care in realisation are essential disciplines in all areas of submarine design. Of particular help in that regard is the setting of weight budgets by the submarine designer for issue to the specialist engineers, as this gives them targets at which to aim and an awareness that there are design penalties for exceeding targets.

For the submarine designer the act of setting weight budgets is closely associated with determining weight margins, which is yet another safeguard and one of such importance that we devote a later section of this chapter to margin policy in order to address the issue for space as well as for weight.

4.9   To finish off this section on weights, we turn to how weight assessment is actually carried out. Even early on in the design process it is possible to make quite accurate estimates of the weight of pressure hull structure, because the scantlings can be closely defined using a relatively small amount of geometric data, by the computer programs available for the purpose. External hull structure and major internal structure are not as amenable to such precise treatment, but workmanlike estimates can be made early on using the weight data base. In that way, a good deal of the structural weight – representing nearly half the total weight – can be estimated without undue risk of gross error. In the same category of robustness from the weight assessment aspect are those large items of machinery and equipment which are likely to be well-defined, like diesel engines, batteries or reactor plant, because the manufacturers will have actually weighed their items. Not quite so straightforward, however, are the weights of the parts necessary to install a particular item in the submarine so that it is supported structurally and connected up to perform its function. It is particularly in the case of systems distributed throughout the submarine that difficulty arises in weight assessment and where the data base previously mentioned – derived as it is from weights recorded in the construction of previous submarines – proves its worth. Allowances have to be made for any changes in geometry in the new submarine design or in the function of its systems, but as long as these were identifiable at the appropriate phase in the design they could be taken into account. There is a real risk that, while major items of weight will be accounted for, some of the many small items associated with the major ones may be missed or inadequately addressed and then accumulate to a significant undercall on what is involved. Even though computer systems enable the vast amount of data entailed to be readily manipulated, they do not in themselves ensure that errors of omission do not occur. There is a vogue for referring to the tendency for the assessed weights of many small parts of a submarine to increase with the passage of time as design progresses as 'weight growth'. Subsequently, in service, submarines like all warships, do in fact grow in weight with time for identifiable reasons, but during design weight growth is a misnomer for what does happen: omitted weight items are found, undercalled weight items are revealed as such and inadequacies in weight control occur. However the phenomenon is described, there is clearly a need both for a weight margin policy and a weight control process to ensure the investment in weight margin is not frittered away.

SPACE

4.10 In examining the space/weight relationship we have already touched on several aspects of space in submarines, when using Tables 4.1 and 4.2 to indicate how and why the densities of the contents of a submarine vary. Our purpose now is to consider space and its utilisation in its own right, before returning to the space/weight relationship in the context of margin policy.

Although in the foregoing we have treated space as if it were just volume, in submarine design, space is also often manifested as a deck requirement area, connected to volume by deck height, but where it is the usable deck area that matters, e.g. for accommodation or control room; while on some occasions space can be manifested with emphasis on a length requirement, e.g. for torpedo stowage. The designer has to change his perspective, from space as a source of volume providing buoyancy support to space as the region inside the three-dimensional envelope of the submarine pressure hull. Within this space the contents have to be disposed to best effect and so the designer dons the mantle of the architect – albeit in a very confined environment – for the shape of the space is significant as well as its amount, and it has to be admitted that in submarines the shape is not all that good for efficient utilisation.

Whereas weight can be treated as independent of shape and its location satisfactorily represented by the position of the centre of mass of the unit, space has to be dealt with at two levels of perception. For hydrostatic purposes it can be treated just the same as weight, i.e. as a quantity located by its centre of volume. However, space has a connectivity characteristic not necessary in weight. If two volumetric demands are adjacent to each other but have different requirements for the geometric distribution of that volume, e.g. differing length, width and height, then what should be a common boundary between them may not be contiguous. There may be left over pockets of space which are not necessarily usable. Where many such 'chunks' of space are put together the total space demand may be greater than the sum of the 'chunks'. Thus the determination of the volumetric size of the hull to contain space demands is not as straightforward as hydrostatics would lead one to expect. As will be seen in a later chapter there are geometric constraints on the form of the enclosing envelope (hull) which can also lead to spatial inconsistencies.

The notion of volumetric efficiency can therefore be introduced which can be defined as:

$$\eta_{Vol} = \frac{\Sigma \text{ Vol demands}}{\text{Total Vol of hull}}$$

It is the art and skill of the designer, as an architect, to try to achieve a volumetric efficiency approaching unity. This presupposes that there is sufficient flexibility in the geometric demands for space to achieve a high degree of interlocking of space to avoid waste. Some space demands will be highly specific, other may only have a few specific dimensions and yet others will be volumetric demands with no specific shape constraints. A typical example of the latter kind is tankage, and so it is possible in the lower half of a submarine to fit in tanks between the more specific demands of batteries and machinery spaces.

In the initial sizing estimate of a new design it is usual to assume, in the interest of keeping size down, that a high level of volumetric efficiency can be achieved. Unused space in a submarine is considered poor design and so the aim is for such pockets as necessarily arise between major space demands to be filled by tanks, storage lockers and system runs. If the designer is unable to make good use of such spaces, it is highly probable that the crew will utilise them to carry extra stores, thereby increasing the weight of the vessel. Thus if a design did have some unutilised space, it would be prudent for the designer to assume a characteristic space density and include it in his weight assessment. The use of space and an appropriate density provide a useful means of preliminary weight assessment in concept design. In fact, concept designs may be developed on a purely volumetric basis with weight estimates based on typical densities.

This method also provides a useful discipline and check on subsequent detailed weight estimations. In essence

$$g \sum_{1}^{n} \rho_i \times V_i = \sum_{1}^{m} w_j$$

for the vessel as a whole. Any serious failure to satisfy this equation should alert the designer to a potential problem. If the left hand side is appreciably greater than the right hand side then significant undercalling or omissions may have occurred in the weight assessment.

MARGIN POLICY AND BUDGETING

4.11 It is natural starting with just a concept when trying to assess the eventual total weight of a vessel of considerable size and complexity – or come to that any artifact with many components – to undercall on the outcome, whether by overlooking some things or not allowing sufficiently for others. We have already discussed ways in which the designer can apply safeguards against undercalling and have commented on weight growth in design as mostly a misrepresentation of the way in which, as a design progresses and more information is acquired about it, (the heuristic process) omissions are uncovered and inadequacies revealed. The particular safeguard on which we now concentrate is the entirely sensible step of allowing margins to cater for such shortcomings, and the associated technique of budgeting.

Because the exercise is one of risk containment – the risk being that of undercalling on all-up weight – the designer will obviously enough base his judgement of what margin to allow for each weight group item on how closely it resembles or repeats previous practice. The issue touches on what we previously termed the evolutionary/revolutionary balance in the new design. Where there is some choice for the designer, it would be prudent for him to avoid an entirely revolutionary design, in which none of its components had been used before in a submarine, because of the many risks to a successful outcome. On the other hand a largely evolutionary design would be unlikely to represent enough progress to justify changing from preceding designs. The judicious choice would, therefore, be a mix of evolutionary and revolutionary in which, it could be argued, the proportion of the former should exceed that of the latter.

In selecting margins for weight group items the designer could afford to set low margins for those of evolutionary character and would be advised to set substantial margins for those which are new or have a considerable degree of novelty. It will be recognised that the term substantial in this context lacks precision, and this is because decision on what is a due amount of margin can only be made by the designer item by item taking account of the particular circumstances relating to each: how much innovation is involved, what the track record of the developer is, whether type testing is to be carried out, and so on. The designer has a particular responsibility in making these decisions and although he would obviously take advice from the specialists involved, it would be politic for him to reserve to himself how much and in what ways he decided to apply margins.

The latter aspect leads directly on to budgeting, an essential concomitant of margin policy, because all contributors to submarine design need guidance on the boundaries (actually upper limits) within which they should work. The setting of budgets is a natural part of good book keeping in what ever activity one might be engaged and this applies to submarine weight control – for it is control that the designer will be aiming to achieve. In fact, weight control procedures extend beyond the design phases into the construction of the submarine, and include the actual weighing of as many components as can realistically be weighed – both as a check on progress in the build-up of weight in the present submarine and as an important contribution to the weight data base for future submarines.

The designer will need to convey to all contributors to the design that weight control is not just a passive tracking system, merely serving to convey when estimates are about to be exceeded, if that were to happen. Thus control must include acceptance by contributors of the need to institute weight saving measures if budgets are likely to be exceeded. It is not possible to generalise about what form such measures might take but clearly, there are limits to how far it would be sensible to go in their pursuit, particularly if there were time or cost penalties (as there might well be) and it is here again that the designer's judgement comes under test.

4.12 This leads to consideration of the role of margin ballast, which we touched on in the previous chapter when discussing the various types of ballast in submarines. Margin ballast is in solid form and is adjustable so that it can be used to adjust weight to buoyancy in the final stages of completion of a new design submarine, and also to provide allowance for future growth in weight. It is not uncommon after the design of a new submarine has commenced, and even when it is under construction, for there to be late changes in operational intent for use of the boat which necessitates changes or additions to the weapons or other combat equipments. It is the policy to set aside some weight margin in the form of adjustable ballast to enable the changes to be made (provided there is sufficient space to do so). Further, because submarines remain in service for many years – 20 to 25 years is quite usual – it is practically certain that some equipments on board, probably weapons, will be superseded by newer equipments entailing some change/increase in weight, which the margin ballast is provided to cater for to a reasonable extent (again provided there is sufficient space to do so). The amount of the margin ballast is a policy decision but it is unlikely to be more than 1% or so of the submerged displacement.

The location as well as the amount of margin ballast is important to its utilisation. Vertically, it is preferrable that this ballast should be stowed high in the boat, e.g. in superstructure casing, so that there would be no impairment of stability when some of the margin is taken up. Longitudinally, the usual choice for location of the margin ballast is close to the longitudinal position of the centres of gravity and buoyancy of the submarine. It should be appreciated, however, that the way in which this ballast might be consumed will depend on where the weight increases for which it is intended to compensate occur. For example, if the propulsion motor experienced an unexpected increase in weight, this would occur in the aftermost part of the boat. Deducting the necessary compensating weight from the margin located near amidships would result in a longitudinal moment imbalance, to counter which the remaining margin ballast would have to be moved forward (Figure 4.2). In extremis, if many individual weight increases occurred in the afterpart of the boat, the residual margin ballast could hypothetically finish up forward of the bow of the submarine.

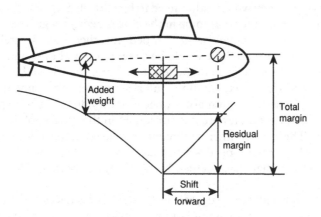

Margin utility to maintain balance

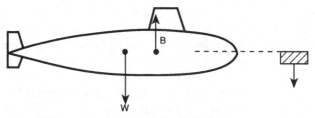

Extreme result of longitudinal imbalance

Fig. 4.2   Margin ballast utilisation

Backtracking for a moment to design and building margins for weight, the designer would assign to individual components appropriate longitudinal and vertical positions depending on where the items whose weight assessment was being safeguarded were located. Although the undivulged totality of the weight margins constitutes an overall allowance for weight increase from any source, they cannot represent an overall allowance for weight increase anywhere in the submarine. The impact would then fall on the permanent ballast, which would be obliged to move, an undesirable but not necessarily dire consequence.

SPACE MARGIN POLICY

4.13 There is an anomaly in the foregoing account, with its apparent preoccupation with possible weight problems, because – as we have postulated – the generality of submarines are space rather than weight driven in design. That raises questions as to whether there should be a space margin policy as well as a weight margin policy and should the two policies be aligned since otherwise, if only a weight margin policy were applied, there would be implied increases in density. Whatever the answers might be in principle, there are practical difficulties in having space margins: they would either have to be parcelled up and distributed around the submarine so as to be in the places being safeguarded against increase in space demands, or lumped together into several larger 'chunks'; if the former, there would be a real risk that they would soon be frittered away by local relaxation, and if the latter, they would fail to provide the intended protection in the intended places. The same difficulty does not arise with weight margins, because they do not require physical manifestation unless they are not consumed in the process of design and building. There is another argument employed against having space margins which might be described as a fear of Parkinson's Law for Space, namely, that space demands will always increase to fill the space available.

Not unexpectedly, we suggest, in these circumstances, in submarine design – because of the high premium on space, illustrated by the relative austerity of accommodation provision – it is not the practice generally to apply overt or covert margins to bids for space. If space bids in a more revolutionary design turned out to have been seriously undercalled as the design progressed, then the submarine would have to be increased in size or, if the discovery of undercalling came too late for that, everyone else amongst the contenders for space would have to squeeze up. In this regard it will be seen that submarines are intolerant vessels; in surface warships there is some flexibility to cope with space undercalls by expanding into the superstructure, but that option does not exist for submarines.

If, when a new submarine design was complete and the first of class boat built, the weight margin had not been entirely consumed, the amount of solid ballast to be stowed on board would be larger than intended. Since the design would generally have been space driven the submarine would not be larger than it need have been, just more stable. If, exceptionally, the design was weight driven the submarine would, in the circumstances being discussed of an incompletely consumed weight margin, be larger than it need have been. Either way, the outcome would be a small penalty to pay for the insurance afforded by weight margin policy against the more serious hazards of undercalling on weight. For to repeat the point we made before, even though the size of most submarines is determined by the space needed for their contents, as design and build progress the submarines appear to all intents and purposes to be weight limited.

OTHER SIZE DETERMINANTS

4.14 To complete this discussion on the weight/space relationship, we return to the matter we touched on earlier, namely consideration of other factors which can affect the picture we have painted for space and weight as ends of the spectrum of determinants of size in submarine design.

We start with what we previously described as the benefits or otherwise of a strictly austere policy for space provision which we call 'shoe horning'. It is not in doubt that if the size of a submarine *is* determined by the space needs of its contents, then to minimise the space provided for all the various functions would keep down its overall size; weight would go down in proportion to keep the average density at that of sea water. There is a long standing tradition in submarine design that to insist on austerity in that regard (as in most regards) is proper – in effect, extending the Schumacher dictum 'small is beautiful' to apply to submarines. Other things being equal, a smaller submarine with the same capability as a larger one should have the operational edge; and there is a temptation to believe that it would be cheaper to build and operate. If correct these would be benefits, so why do we qualify with the phrase 'other things being equal'? The reason is that a submarine is palpably the most dense and complex of marine vehicles, and this is reflected by the high labour costs involved in fitting them out under very confined conditions; to go further in squeezing up on the contents could become counterproductive and almost certainly push up building costs rather than reduce them, as well as making maintenance in service and the work of refitting more difficult and probably more costly. It is for these reasons that it is usual during the submarine design process to make the investment in large scale models and full scale mock ups as well as extensive use of computer aided design graphics. The aim is not to produce the smallest possible submarine for the allotted tasks, but one which represents a good compromise between operational effectiveness and least through life cost overall; that aim is then the size determinant.

Equipment location

4.15 It is implicit in much of what we have said in this chapter that the majority of the equipment deployed by a submarine should be contained within the envelope of the pressure hull. This is open to question because in some submersibles developed for commercial use the pressure hull is a sphere – splendid for strength purposes but appalling for space utilisation, and necessitating locating as much equipment as possible outside the pressure hull. Now it is a well-established principle in submarine design – based on bitter experience – that no item should be exposed to sea water at diving pressure unless there is compelling reason to do so. However, anticipating the factor we come to next, the use of a wrapround external hull, if there is free-flooding space outside the pressure hull that cannot otherwise be put to good use, then it would be sensible to utilise it for reducing demands on space inside the pressure hull. This would allow some reduction in the size of the pressure hull which, while it might not result in an overall size reduction, could contribute to limiting the size increase that would otherwise occur. The scope for size reduction or limitation in that way is not readily apparent; for example, if the liquids usually stowed in tanks inside the pressure hull in present practice were stowed outside, there would be a problem in making use of the poor quality space so made vacant. Nevertheless, if the incentive were there, the potential contribution of this alternative approach to size determination could be established.

Wrapround external hull

4.16 The arrangement usually adopted for the external space of *Albacore*-derived submarines, which we described in Chapter 3 on Submarine Hydrostatics, leads to its disposition at the fore and after ends of what is otherwise bare pressure hull, and this we described as a single hull configuration. It is recognisable that the exposure of the bare pressure hull is potentially a source of weakness under weapon attack. This could to some extent be ameliorated if the external hull arrangement once common in other ocean going diesel electric submarines – the double hull configuration – were adopted. The annular distance between the external and pressure hulls would afford some stand-off against weapon attack which would reduce if not avoid the hazards with no stand-off at all. Inescapably, a wrapround external hull would cause a considerable increase in structural weight and in the form displacement, if not in submerged displacement of the submarine. (What happens to submerged displacement would depend on how much of the annular space between external and pressure hulls was used for main ballast tanks, fuel stowage and other tanks or just left to be free-flooding). More than that, the submarine would be pushed to the weight driven end of the spectrum of size determination previously mooted. No longer could the pressure hull be used for fine tuning of weight/buoyancy balance by adjustment of the deep diving depth requirement – unless there was a change of the material used for the pressure hull and frames from steel to one of low density, which is the next and last factor we go on to consider. Before leaving the issue of the wrapround external hull, however, it is worth observing that the annular distance between external and pressure hulls needs to be of the order of 1m for access during construction and maintenance and this represents a considerable increase in enveloped volume, while contributing only a very modest stand-off.

Pressure hull structural materials

4.17   To conclude this treatment of the weight/space relationship we touch briefly on what the effect on submarine size might be of changing from steel for the pressure hull plating and frames to a less dense material. We discuss the issue of choice of material for pressure hull structure in Chapter 5 on submarine structural design and for now confine comment to the statement that possible alternative materials are titanium and fibre reinforced plastic, FRP, (aluminium being unsuitable for military submarine purposes). Titanium has a Young's modulus and proof stress higher than many steels, can be welded, is corrosion resistant and very expensive. FRP has a lower Young's modulus and proof stress than steel, has to be bonded, is corrosion resistant and not unduly expensive. (GRP, glass reinforced plastic, is already in use for submarine external structure to a limited extent).

Briefly, and as can be inferred from what we have said earlier in this chapter, there may be no advantage in terms of submarine size in using a lighter material for the pressure hull of a submarine which is space driven – as we maintain the majority of single hull submarines are. The lower density could be used to increase deep diving depth while investing the same weight in pressure hull structure as with steel, though it is important to remember that all systems and tanks inside the submarine which are subject to sea pressure would also have to be increased in strength – and weight – to match the pressure hull. There might be other worthwhile advantages in changing from steel, like reducing the magnetic signature or changing the acoustic properties of the submarine, but in the present context of size determination, it would not be worth changing.

That leaves the option of changing to a lighter material for pressure hull structure as a means of compensating for the increase in total structural weight which would otherwise occur on adopting a wrapround hull configuration, but this would seem to be more a matter of minimising penalty than of achieving benefit.

The strongest case for adopting other hull materials arises if there is an imperative operational requirement for deep diving depth. In that case the design option of adjusting diving depth capability to achieve weight/buoyancy balance may not be available and the design may become weight governed with a steel pressure hull. Rather than increasing the size of the vessel and creating under utilised space it may in those circumstances, be better to adopt an alternative hull material which restores the balance.

REVIEW

Whilst there are many complex arguments in the determination of size we commend that, in the interests of keeping the size of submarine down, the initial sizing should be based on the space demands of the vessel. With a prudent margin policy the weights should be controlled to achieve overall neutral buoyancy, i.e. balance. Only if there were over-riding requirements should a weight determined sizing be adopted.

5 SUBMARINE STRUCTURES

INTRODUCTION

5.1 The main attribute of a submarine is its ability to dive beneath the surface and to go to reasonable operating depths. For a manned submersible there is a requirement for the enclosed volume to be maintained at atmospheric pressure. This need applies not only for the personnel but also for much of the equipment which has been designed to operate in atmospheric conditions. It is desirable to keep the enclosed volume as small as possible so as to limit the weight of structure that is required to withstand the differential pressure between sea pressure at depth and atmosphere. In small unmanned submersibles and ROVs it will usually be possible to minimise the amount of volume that needs to be contained by the pressure carrying structure, but for most large seagoing manned vessels there are inescapable requirements for a considerable amount of volume to be contained within the structural envelope. It may be that the design of a submarine as a whole leads to a decision to include other volumes within the pressure hull although they are not necessarily required to be at atmospheric pressure; for example, some of the main ballast tankage may be included within the pressure hull. It can also be convenient to locate some fuel tanks within the pressure hull so that they can be operated in atmospheric conditions. Variable ballast tanks will usually be located within the pressure hull, even though in some instances they are subject to sea pressure. It is important that where such volumes are inside the pressure hull, steps are taken to enable them to be isolated from sea pressure.

Before we describe the technical aspects of submarine pressure hull design, we set the scene by discussing the operational considerations that lead to the depth requirement.

OPERATIONAL REQUIREMENTS FOR DEPTH

5.2 As with many other vehicular characteristics the depth requirement whilst occasionally set by very specific considerations is, more often than not, subjective. There is an understandable tendency for the view to be taken by operators that the larger the better. However, seeking to operate at considerable depth will incur a cost penalty as we have already indicated. The submarine design changes from volume controlled to weight controlled and this leads to a larger vessel with consequent increases in all other vehicular demands, e.g. power for same speed. This argument implies that depth capability is an outcome of design rather than an input, which could be regarded as an unsatisfactory feature from the operational viewpoint. Before looking for more rational bases for deep diving depth the question must be asked: would it be put to good use? With many conventional submarines virtually the only experience of full diving depth occurs during test dives. (The reasons for this will be discussed later.)

One approach to answering the question is to examine what determines the minimum satisfactory depth capability: the submarine must submerge but by how much? Clearly the absolute minimum is such that all of the boat, including the bridge fin, is submerged so that it is at least, laterally, out of sight. It may be still visible from directly overhead but that would depend on the translucency of the water. Such a shallow submergence would put the bottom of the hull 20 metres or so below the surface and the boat would still be subject to surface conditions, e.g. storm waves, and could find it difficult to remain submerged. It would also be subject to wavemaking resistance as in the snorting condition and the surface waves so created may be detectable. Perhaps a more important factor is that at such a shallow submergence the submarine would be vulnerable to collision with surface ships which could be unable to see and may not hear. Deep laden supertankers may have a draught of 25–30 metres and so for insurance submarines in general should have a minimum operating depth of say 50 metres. As we show later, safety aspects dictate a design collapse depth considerably more than that.

5.3 Having established a tentative minimum, what operational factors lead to seeking deeper diving depths? A naval submarine will need the capability to avoid detection and evade attack from surface and air units. The primary form of detection is by acoustics, very often using the active mode from surface units. Sound waves rarely travel in straight lines in sea water, the variations in temperature and salinity causing their refraction. This phenomenon can often result in a surface duct in which sound waves initiated at the surface are refracted back up to the surface along a curved path of only limited depth penetration. The surface duct depth will vary depending on ocean climatic conditions, but it is clearly advantageous to a submarine to be able to get below this depth where it can avoid detection from the surface surveillance. It is not our present purpose here to set a figure to such a depth, just to observe that it is a factor in the argument for deeper diving. In the evasion mode a large depth range capability for a submarine poses an additional problem to the attacker as, to be effective, his weapons must have the correct depth setting or, if it is an autonomous search weapon, the volume and dimensions of the search pattern are considerably increased.

The variation in the density/temperature structure of the ocean can result in other layers and sound channels which a submarine can exploit for evasion or for its own surveillance purposes. Some of these are hundreds of metres deep and so the operators' view that the deeper the better has some justification.

World War II submarines had a depth capability of 100–150 metres. The majority of modern submarines have a depth capability of about 3 times these figures. This order of depth can usually be achieved by a volume dominated design using high yield steel as a structural material.

5.4   An entirely different argument for a reasonably large depth capability stems from the control dynamics of a submarine addressed in Chapter 8. For now it should be recognised that for safety reasons a high speed submarine needs manoeuvring room in depth. An error or fault in control at high speed could lead to a large excursion in depth before recovery action could take effect. This may result in an inadvertent surfacing of the boat, which would be embarrassing though not necessarily catastrophic, or a rapid descent to below collapse depth which most certainly would be catastrophic. Hence a very significant factor in determining minimum collapse depth could lie in the fault analysis of a submarine's high speed control systems.

The foregoing debate is based on military submarine aspects and does not take into account the relatively easy decision on design depth of an exploration or search vehicle which has a specific requirement to go to the sea-bed in its area of operation.

5.5   The question was posed earlier of how often a deep depth capability would be likely to be used. Many submarine operations involving blockade or surveillance take place in operational areas of Continental Shelf depth or even shallower parts of the ocean. In those circumstances the sea bed sets a limit on the depth of operations and greater diving depth cannot be exploited. Even in the deeper ocean the conventional submarine is usually tied to near–surface operations, e.g. to use its periscope, radar and communications systems and to snort to re–charge its batteries. Although depth capability is useful for evasion, there is little other reason for going very deep and having to return to the surface at regular intervals. Thus many boats, even with a reasonable depth capability, do not use this feature very often. The high powered, high speed nuclear submarine with considerable sustained submerged endurance can exploit depth performance to a greater extent. However, as mentioned, at full speed its operators must prudently leave a goodly depth margin for safety. Even the nuclear boat has to come near surface to use some of its sensors and communications. Thus, in general, few submarines make frequent use of a deep depth capability. As will be seen later, for material reasons frequent excursions in depth can affect the life of the pressure hull and so that activity has to be discouraged and limited to operational necessity. Hence, in view of the cost implications compared to the utility, many submarine designs are accepted with moderate depth capability.

We now go on to review some of the technical features involved in pressure hull design. It is not our intention to provide detailed methods of structural design, which is a specialist subject. The object rather is to show how such considerations impact on the overall design of a submarine and constrain some of the freedoms that might otherwise appear available. As an example, in light of the arguments in Chapter 4, why not adopt square pressure hulls which would be more efficient volumetrically?

SHAPE OF THE PRESSURE VESSEL

5.6   As we showed in Chapter 4, the pressure hull structure constitutes a large proportion of the total weight of a submarine. It is therefore essential that this structure should be as efficient as possible in order to limit its weight. (For very deep diving the hull might even weigh more than the buoyancy it displaces, necessitating other sources of buoyancy). Efficiency means utilising the strength of the material to its maximum extent throughout the mass of the material. For practical purposes the pressure differential to which the structure is subjected can be considered as uniform; for although pressure increases with depth and the hull has a finite depth of its own, the difference between the pressure at the top of the hull and the bottom of the hull is, in most cases, relatively small. However, it should be appreciated that if the hull is, say, 10 metres diameter, the differential hydrostatic pressure between top and bottom is equivalent to one atmosphere.

5.7   For uniform pressure the ideal pressure vessel geometry is a sphere (Figure 5.1 (a)). A thin shelled sphere, subject to uniform differential pressure between the inside and the outside, will experience equal strains and stresses throughout all the material of the shell. It is the natural lowest energy state taken up by a thin membrane, thin bubble film or balloon if subject to an internal pressure higher than the external. Although, for a submarine, the pressure differential is of the opposite sign, i.e. pressure external higher than internal, the same principle of uniform stress and strain applies. For some small submersibles a spherical pressure hull is adopted for that reason. However, it has its drawbacks: it is not an ideal shape when considering the hydrodynamics of the vessel at speed, which calls for an elongated streamlined form. A sphere also poses difficulties in filling the space usefully: it is all curves and if one considers the largest cube that can be incorporated within a sphere there is a considerable amount of unused volume. It is also difficult to manufacture, having curvature in both directions so that plates cannot be rolled to shape. To form a sphere the plates have to be pressed, or spun as two hemispheres, and then welded together.

5.8 Next to a sphere the most efficient pressure hull structure is that of a right circular cylinder with domed enclosures at either end (Figure 5.1 (b)). This is a commonly adopted geometry for pressure vessels intended to contain pressure. It gives the geometric freedom to vary the diameter to length ratio and therefore it is more readily possible to fit a cylinder within a streamlined form. It is slightly better in terms of volumetric efficiency, in that more rectangular space can be included within it than a sphere of the same enclosed total volume. It is simpler to manufacture because the cylindrical hull has single curvature and so plates can be rolled. It is not quite so structurally efficient in that there is now a two to one stress relationship as between the circumferential and the longitudinal directions in the material.

Even the right circular cylinder with dome enclosures is not an ideal shape from a hydrodynamic view point. That requires a varying diameter along the length, rather fuller at the fore end and tapering towards the after end. If the submarine design requires that the pressure hull should form the major part of the envelope, then it has to conform to this shape, which can be accomplished by a combination of cones and cylindrical sections each of which can be rolled in a single curvature and welded together to form a more streamlined body enclosed by domes of a smaller diameter at the ends (Figure 5.1 (c)).

Except for very deep diving vessels the pressure structure is relatively thin and therefore relationships to membrane analysis can be drawn.

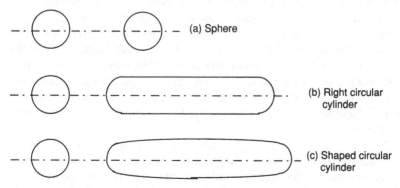

(a) Sphere

(b) Right circular cylinder

(c) Shaped circular cylinder

Fig. 5.1   Ideal forms

Though, close up, the structure of a submarine may appear massive, in relation to its diameter it is relatively thin and akin to the skin of a balloon or the paper on a cigarette. It has much in common with the modern technology of pressurised containers such as beer cans and gas cylinders, but has to withstand external pressure rather than internal pressure.

5.9   The foregoing discussion of the form of the pressure hull relates to most modern submarines with axial propulsors and a basically axial streamlined form. In earlier submarines of this century, the style was of a pressure hull within an external hull of boat-shaped form and with raked twin shafts below the stern. For such vessels it was common practice for the pressure hull to be so configured that the top line was straight and level. The central circular cylinder part was connected at each end to canted conical sections which rose up at either end (Figure 5.2(a)). The conical sections were closed off either by small domes or by massive single or double flat bulkheads, the space between double bulkheads forming part of the tankage requirement. Some boats departed from circular cross-section as well, the cone at the after end having a flattened horizontal oval shape suited to the twin shaft propulsion arrangement; while the fore-end cone was of vertical oval section suited to two vertical rows of torpedo tubes in the weapon discharge arrangements. It will be evident that such structural configurations depart from the concept of membrane containment of differential pressure, because non–circular sections will be at all times subject to ring bending moments requiring much heavier framing. One reason for the older submarines being able to adopt such unfavourable structural configurations was that they had a relatively shallow diving depth, usually less than 100 metres, and so it was possible to accept the greater structural weight of such a design.

Another older submarine design departed even further from the circular cross-section, adopting a figure of eight shaped cross-section (Figure 5.2(b)). However, this shape is not actually such a gross departure from the membrane concept. Essentially the cross-section is a combination of upper and lower circular hulls merged at a horizontal segment, the missing parts of the circular shells being replaced by a horizontal deck. That geometry can be observed to occur naturally in merged soap bubbles; the individual bubbles form spherical membranes but if two or more are partially merged or connected the connecting boundary forms a flat membrane between the spheres. It can be shown mathematically that this merged membrane configuration is the lowest energy state of the combined membrane system.

Thus, although earlier in this section we advised circular sections as the usually appropriate hull configuration, designers need not be inhibited in their hull design if the overall concept warrants a change. It is conceivable that two or more cylindrical or spherical hulls could be merged if greater width or depth of the hull was required, provided that particular care was taken in the design of the junctions and the internal supporting bulkheads or pillars. But a word of warning to any designer contemplating such a novel configuration: there has been considerable research and investigation of the stress distribution and failure modes of the ring stiffened cylinder, cones and dome bulkheads; to make the change would call for a considerable investment in research and structural evaluation before it could be adopted with confidence as to its overall safety.

ELASTIC DEFORMATIONS OF THE SHELL

5.10 Though much structural analysis concentrates on the stresses experienced by the material related to its yield or ultimate strength, for an understanding of behaviour it is important also to take account of the strain or deformation under load. Thus if we consider first the shell of an unstiffened sphere subject to a differential pressure, it will change diameter and be uniformly strained in all directions. Under internal pressure a sphere will in consequence get slightly larger in all directions while, in principle, under external pressure the opposite will happen and the sphere will uniformly shrink in its diameter. Considering now the shell of an unstiffened cylinder of circular section under differential pressure, it will experience different strains in the circumferential and the longitudinal directions. However, as an approximation it can be taken that all along its length the diameter will change equally with the differential pressure. Since an open-ended cylinder cannot contain differential pressure the ends must be closed with some form of dome. If these are hemispherical domes the strain relationship at the junction between the cylinder and the sphere must be considered. Regarded as unconnected elements initially, there would be a uniform diametric strain of the cylinder which may be

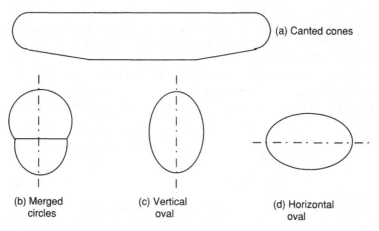

(a) Canted cones

(b) Merged circles   (c) Vertical oval   (d) Horizontal oval

Fig. 5.2   Older forms

twice as much as that of the hemisphere to which it is to be joined. When the two elements are joined, junction strains and stresses will occur because of the mis-match of the strains between the elements. Considering a longitudinal cross-section of the junction, (Figure 5.3 (a)), it will be seen that in order for the elements to remain connected together there has to be a curvature of the shell longitudinally to effect the change from the strain of the cylinder to that of sphere. Thus the connection of the two structural elements introduces strains and hence stresses which might be of large magnitude compared to the free shell.

5.11   A similar situation will arise where a cylinder is connected to a conical element. It is less easy to envisage the distortion of a conical section, but nevertheless under external pressure it will tend to shrink and its conical shape will provide some stiffness against the shrinkage. If the larger diameter end of the cone is connected to a cylinder, there will be an inward shrinkage under external pressure of both the cylinder and the cone which may not be compatible. (Figure 5.3 (b)) In addition there will be the thrust along the axis from the end pressure component. This thrust, at the junction, provides an outward strain tendency in opposition to the inward strain of the cylindrical and conical shells. Thus, particularly sharp changes of curvature of the material can occur in the locality of the junction, which can lead to very high longitudinal bending stresses which are difficult to reduce to acceptable levels. For a cone connected at its small diameter end to a cylinder a similar situation occurs but the end thrust acts to push the junction inward, which is not in opposition to the general shrinkage of both the cone and cylinder. Whilst some local bending stresses can be expected at such a junction, they are not of the severity experienced in the external cone/cylinder junction. It is therefore important in the design of a pressure hull to avoid such sharp junction stresses if at all possible.

However, in the design of a submarine it is not uncommon to encounter situations in which there is a requirement for the pressure hull to be of a large diameter over a part of the length of the vessel and of a smaller diameter over another part. If the two different diameters are simply stepped together there would need to be a flat circumferential bulkhead between the two, which would have to withstand external pres-

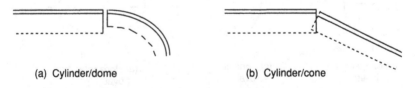

(a) Cylinder/dome   (b) Cylinder/cone

Fig. 5.3   Incompatibility of deformation under pressure

sure and the non-aligned thrusts of the two cylinders. A structure like that would be extremely difficult to design so as to keep within material limits. If a cone transition were devised to link the two together, this could lead to the high stresses and strains at the junction that have been described. One way of imagining how to minimise the effect is to consider the shell to be made of thin rubber like a balloon of the appropriate geometry and visualise what would then happen under internal pressure. At the large diameter cylinder/cone intersection the cylinder would expand; so would the cone and it would tend to bulge outwards, becoming more spherical in this area. At the small diameter end there would be less tendency to behave in that way, both the small cylinder and the small end of the cone expanding equally together. As such a membrane cannot support bending stresses, the conclusion is that the shape which would avoid bending stresses is one in which the large diameter cylinder is connected by a spherical transition piece into the cone before connection to the small diameter cylinder. Although that configuration would be more costly in construction terms it would provide a more even stress/strain distribution in the pressure hull. Nevertheless, the additional cost and complexity of the structure suggests that the designer should avoid such a configuration if alternative solutions to the space needs of the submarine design can be found.

BUCKLING DEFORMATIONS OF SHELLS

5.12 Whereas it is possible to support internal differential pressure by a thin unstiffened cylindrical shell as has been described, when the differ-

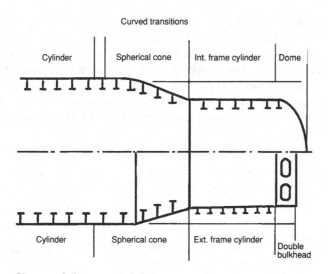

Fig. 5.4   Change of diameter and closure

ential pressure is predominantly external, that structure will not be able to support pressure up to the limits of its yield strength. There is an analogy with a thin rod subject to end loading: if the end loading is tensile then the strut can be fairly thin and will take all the stress up to its ultimate strength. However, if the load is reversed to compression then, long before ultimate material strength can be reached, the strut will buckle out of the way and collapse unconstrainedly, without being able to sustain any increase in pressure beyond that at which collapse occurred. A similar situation arises when external pressure is applied to a vessel. A short length of unstiffened pressure hull cylinder can be regarded as a circular strut (Figure 5.5).

At each point around the circumference there is compressive loading applied between the elements of the circumference. Any short length of circumferential material may be liable to buckle either outwards or inwards, depending on any small deviations that it may have had from its original circularity. The circumferential hoop can be regarded as an infinitely variable length strut and so there are many mode shapes in which it may buckle. The most probable mode of failure of an unstiffened cylinder is that in which it distorts into an oval shape and eventually flattens under the pressure. The failure pressure will be well below the inherent strength of the material, which cannot develop its full potential to withstand loading. It is therefore necessary to support the shell against premature elastic buckling by the provision of stiffening to hold the circular shape. The usual method of stiffening is by use of ring stiffeners fairly closely spaced along the length of the cylinder, thereby providing radial stiffness to the cylinder and inhibiting the lower modes of collapse.

5.13   In providing radial stiffness the ring stiffeners also increase the complexity of the stress and strain experienced by the pressure hull shell. The strain pattern is of variable circumferential strain along the length. The strain will be lower over a stiffener and a maximum half way between one stiffener and the next, and this results in bending strains in a longitudinal direction, with bending in one sense over the stiffeners and reverse bending between the stiffeners. Consequently, in way of the

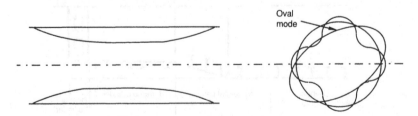

Fig. 5.5   Modal buckling of unstiffened cylinder

stiffeners the shell plating will, at its outer fibres, experience circumferential stress reduced by the support of the stiffener and a longitudinal strain resulting from the combination of tensile bending and endload compression. At the innermost fibres of the shell at this location, the bending strain over the stiffener will add a compressive strain to that from axial load, giving a two-way stress situation which may cause plastic deformation and so precipitate failure of the material.

Between stiffeners there is an increased circumferential strain relative to that over the stiffeners. Longitudinally there will again be a combination of strains due to axial loading and bending which results in a larger compressive strain at the external fibres of the material and a reduced strain on the inner surface. The curvature there will be less than that over the stiffener. Against this effect, the end loading on the cylinder acts on a changed radial position of the plating which causes an additional bending strain, rather like that experienced by a buckled strut. As a result it is possible for a ring stiffened cylinder to fail in what might be described as a concertina mode, by the material folding up in between stiffeners. An example of partial concertina folding is shown in Figure 5.6(a), where a model has been subject to external pressure and the pressure released before it totally collapsed. This mode of failure is known as interframe collapse and as described later is the preferred way for a pressure hull ultimately to fail as it is relatively predictable.

Though it is possible to compute the stresses and strains occurring in the complex distortion pattern described above it is found in practice that small deviations in geometry and material properties can cause this mode of failure to occur at different pressures. Recourse is therefore taken to physical tests of models of typical pressure hull geometry and materials to establish a relationship between theory and reality.

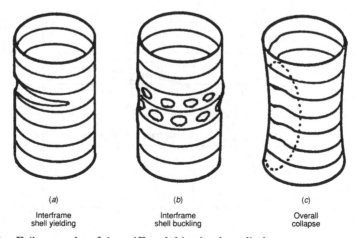

(a)
Interframe
shell yielding

(b)
Interframe
shell buckling

(c)
Overall
collapse

Fig. 5.6   Failure modes of ring stiffened thin circular cylinders

Figure 5.7 shows experimental results plotted against parameters described in more detail in Appendix 4. It is a matter of design judgement whether to adopt an ultrasafe or an average line through the data, but this approach does provide reliable solutions. The need to resort to empiricism and the dependency on statistical data rather than individual results reinforces our earlier point that considerable research and testing are required before a novel structural configuration or new materials could be adopted with confidence.

5.14 A ring stiffened cylinder of very long length can be treated as if it were an unstiffened shell, even though it has a greater stiffness to radial deformation. There remains the possibility that a long combined stiffener/shell plating cylinder might collapse in an overall buckling mode and become flattened. To provide additional radial stiffness and maintain circularity against overall buckling modes it is necessary to introduce occasional heavy stiffeners or occasional complete transverse bulkheads. The spacing of these bulkheads is determined by keeping the length between them short enough to avoid the risk of overall buckling failure. Inevitably this spacing interacts with the desirable layout of the internals of the hull. The additional radial stiffness provided by a bulkhead adds to the strain situation in the shell plating where it passes over this particularly stiff plane by increasing the longitudinal bending strain in the shell. It may in consequence be desirable to increase the thickness of the shell locally by way of a bulkhead and to ease the bending stresses by closing up the ring stiffener spacing adjacent to the bulkhead.

Structural design computer programs exist for the analysis of ring stiffened cylinders and the junction stresses (Figures 5.8). They can also

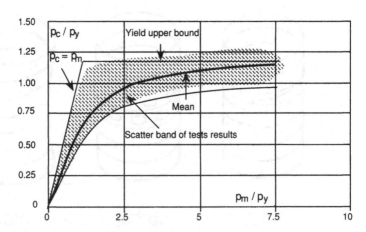

Fig. 5.7   Interframe shell collapse results

provide the designer with guidance on the optimum selection of shell plating/stiffener geometry and spacing to obtain minimum weight of the primary hull structure. However, because the stiffeners constitute an interference with the space inside the hull, overall design considerations may justify use of a non-optimum spacing of stiffeners and acceptance of a small weight penalty.

OTHER FAILURE MODES

5.15   Whilst the ring stiffeners are introduced to prevent premature buckling of a cylindrical shell they are themselves subject to buckling modes. They are effectively circular struts, albeit of greater cross-sectional area and moment of inertia than the shell plating. In conjunction with part of the shell plating a stiffener may buckle out of the cross-sectional plane through the stiffener. It is also possible for the ring stiffener to buckle out of plane in a torsional mode. Thus the introduction of stiffeners whether as rings or as bulkheads to prevent forms of shell failure not only complicates the stress pattern within the shell plating but also

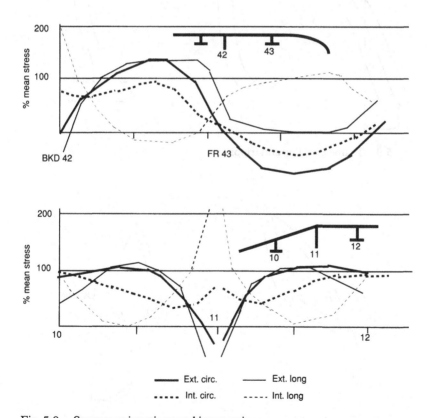

Fig. 5.8   Stresses at junctions and intersections

increases the number of ways in which the structure as a whole may fail (Figure 5.9).

The design of a pressure hull structure therefore involves considering the structure as a whole as well as each of its constituent parts with the aim of ensuring that no single detail would be liable to fail prematurely before the rest of the structure. As a design philosophy the object should be for the material to at least reach its yield stress before failure. An essential aim is to avoid any premature failures due to elastic buckling, and moreover that the pressures to cause any form of buckling should be well in excess of pressures resulting in yield failure. As is the case with all compressive buckling failures, the geometry of the structure is extremely important to the initiation of buckling. If it were geometrically perfect there would be a possibility of reaching a

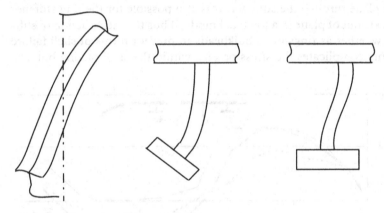

Fig. 5.9   Frame buckling and tripping

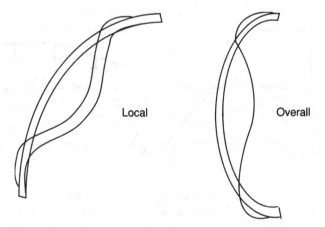

Local   Overall

Fig. 5.10   Dome buckling

very high pressure before buckling was initiated. However, any geometric errors, such as departures from circularity or departures from straightness of webs of the frames, can lead to precipitation of failure. Design of pressure hull structure has to take account of such imperfections, and analytical methods and computer programs are available for that purpose.

The dome closures at the end of the pressure hull also pose buckling problems. It is less easy to visualise the modes in which a dome may fail. The most likely mode is a total implosion inverting the dome over the end of the cylinder. (Figure 5.10) Geometric accuracy has a great influence on the buckling of the dome. If a small area were not at the right curvature this could prematurely collapse, initiating overall collapse of the dome. The domes are not necessarily entirely unsupported, as they may have an egg crate structure welded to them to support external structure or tankage and this will help to support the dome against buckling failure. At the same time these stiff points lead to local bending and stressing of the shell.

INTERNAL SUPPORTING STRUCTURE

5.16   Within the pressure hull the volume will be subdivided not only by the transverse bulkheads which perform the function of supporting the cylinder but also by decks and, in the lower part of the hull, tank structures. It is usual to ignore the supporting effects of the decks and tank structures on the pressure hull, i.e in the design of the pressure hull it is assumed to be empty and not have any structure within it apart from the main transverse bulkheads. However, decks, if welded directly to the pressure hull, have the effect of acting as ties or props across diameters or sectors of the hull, inhibiting displacements at these locations. Thus, for example, a horizontal deck at mid height of the cylinder, will tie the cylinder across the middle and therefore inhibit the mode of collapse in which the top and bottom of the hull approach each other and the sides go out; the deck would tend to favour an angular mode of failure where it lies along a line which does not change length with the buckling mode. (Figure 5.11) Decks are not always

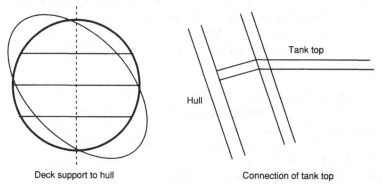

Deck support to hull   Connection of tank top

Fig. 5.11 Internal support to hull

connected solidly structurally to the pressure hull, however, as to do so imposes loads on the deck as well as on the cylinder, which could lead, as the pressure hull contracts under pressure, to a compressive load on the deck causing it to buckle.

It has been known for decks fixed in that way to buckle at some depth in an elastic manner and straighten up as the vessel comes to a shallower depth. Because of these undesirable effects it is sometimes the practice to have 'floating' decks, i.e. decks supported from the ring frames but in such a way that they do not take a direct compressive load from the structure, which can be achieved either by 'dog-legging' the edges of the deck or by suspending the edges. This isolation of the deck from the hull can serve a useful additional purpose by isolating vibratory or noise-making equipment mounted on the deck from direct structural connectivity with the sea. A lower deck which may act also as a tank top to the bottom sector of the cylinder is not so easily isolated from this structure, particularly if the tank has to withstand sea pressure. Tank structure of that sort has to act as an integral part of the pressure hull structure and so can influence the failure modes of the cylinder by stiffening that part of the sector of the circle. It is normal practice to assess the out-of-circularity of a pressure hull cylinder by measurement of circularity of the shell plating, but the true circularity is that of the neutral axis of the ring structure i.e. shell plating plus framing plus any tank frames or webs that may be involved. If those structural components are taken into account then, in the lower part of the hull, the neutral axis will depart considerably from circularity. At the same time, however, there is a massive increase in the moment of inertia of this part of the hull due to the tank top and webs, so that departure from circularity has not been considered to be a significant influence on the buckling modes of the pressure hull. Though it is not possible to isolate the tank top structure from the pressure hull it is considered good practice to avoid the awkward angle of intersection of the tank structure to the hull which occurs if the tank top is kept in a horizontal plane. A suitable alternative is to bend down the edges of the tank to intersect the hull shell plating at right angles. (Figure 5.11) That configuration provides for better welding at the joints and also some flexibility across the tank top so that it is less directly loaded by the compression of the hull.

PRESSURE HULL PENETRATIONS

5.17   Though the pressure hull is intended to be a pressure tight envelope containing atmospheric pressure within an enclosed volume, it is necessary to provide means to pass from inside to outside, both whilst submerged and more particularly when on the surface. For that purpose hatches are provided for access of personnel and stores and penetrations for pipes, cables and shafting connecting the various equipments sited external to the pressure hull but controlled or actuated from within. Examination of an expanded shell drawing of a typical submarine's pressure hull with all the penetrations indicated is very revealing. Despite having given careful effort to the design of the pressure tight shell, in many places the designer has had to accept it being disrupted by penetrations. Care has to be taken in the design of the penetrations as each constitutes a source of initiation of failure of the pressure hull. It is usual for most penetrations to be circular, though on occasions a long elliptical hole cannot be avoided. A circular hole in the plating results in a stress concentration at the edge of the hole. For a flat plate with the two-to-one stress system, characteristic of pressure hull plating, the stress concentration factor at the hole is 2.5. For a curved shell it is possible that instead of the strains remaining in-plane at the edges of the hole, the plating may move radially and lead to higher stress concentrations than with a flat plate. However, it is usual to form such a penetration with a cylindrical support through the hull plating, or in the case of large openings a coaming around the edge of the plating on which to place the hatch. Whilst this type of insert is usually designed in order to accommodate the hatch fitting, it also performs a structural function by supporting the edges of the curved shell and preventing the radial distortions of the hull. It is then reasonable to assume that plane stress concentrations will arise even in a curved shell. (Figure 5.12)

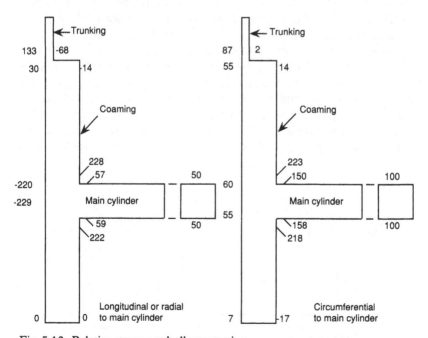

Fig. 5.12 Relative stresses at hull penetration

5.18 The largest openings in the hull are normally those associated with the access hatches at the top of hull which are sized to enable personnel and stores to pass through. (Figure 5.13) That size will usually be greater than the ring stiffener spacing of the pressure hull. It is therefore necessary to cut ring frames in order to make such hull penetrations. It has been the practice to cut the frames in way of an access hatch and butt them to the coaming of the hatch, but it must be remembered that as these frames carry compressive load around the circumference of the pressure hull, cutting them weakens their support to the pressure hull. One way of overcoming this weakening effect would be to insert compressive pieces between the cut frames when the hatch was not in use. At depth it would be apparent that these props do take the load because they would become very tight in place. However, this is not always a practical solution, and so consideration should be given to enabling the load from cut frames to be shed sideways by bifurcation to adjacent frames on either side of the hole, and then reinforcing these adjacent frames to carry the additional load transferred from the cut frames.

Most major penetrations in the hull are designed individually but where there is an area in which a number of smaller penetrations occur in close proximity it is common practice to thicken up the shell plating over that area, perhaps even doubling it to provide a lower stress zone within which the holes can be made without other reinforcement. In that way, a multiplicity of close welding for individual insert plates in the shell to provide seats for glands and hull valves can be avoided. Another common practice, by which a shell seating pad is provided to which a valve will be bolted or studded, has the drawback that the pads can give rise to stress raisers. A way now adopted for some major penetrations is to provide a forging which acts as both the support to a penetration and as part of the shell plating, and which can be welded into the shell plating as an integral part of the pressure hull. This alternative reduces the amount of welding on the hull and reduces the number of stress raisers; it also provides for better continuity of the structure of the shell around the hole.

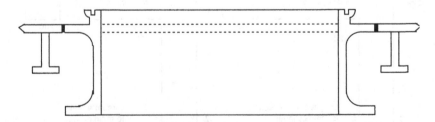

Fig. 5.13 Example of large hull opening

5.19   Many penetrations in the hull are for the purpose of piping sea water to various services within the hull. These pipes have to be designed to withstand full sea pressure and so may be of thick walls and quite stiff. Since the pressure hull contracts under pressure this movement imposes a displacement on the hull/pipe connections, which can result in high strains within the pipes making them prone to failure. Care has therefore to be taken to provide flexibility in the piping runs, either by means of tortuous indirect runs or by introducing bellows or flexible elements. Whilst this is a matter of detailed design, account has to be taken in overall design of the additional space that such arrangements take up.

The penetrations in a pressure hull are the most likely source of flooding at depth. The failure of a major penetration in the hull would almost certainly result in the vessel flooding and perhaps of being unable to reach the surface. If failure occurred actually at a hole there would be no way in which its consequences could be avoided. Where a large number of small penetrations are required through the hull it is for consideration that they should be incorporated into a pressure vessel within the pressure hull and connected to sea by a single major, carefully designed, penetration through the pressure hull with an isolating valve for closure in the event of an internal failure.

EFFECTS OF SHOCK

5.20   The penetrations in the hull are also the most likely cause of failure under shock conditions from standoff explosions. The pressure hull itself, although quite stiff under slowly varying loads, is relatively flexible under very fast shock loads and so can withstand quite large explosions. Penetrations and hatches are, however, local hard spots in this generally flexible structure and that can cause high stresses and failures at such locations. The most likely failure mode of the pressure hull as a whole is an overall whipping response, because the structure being designed primarily for circumferential stress is not longitudinally stiffened. If the explosion induced a resonant vibration of this long circular sectioned beam, failure could result. It is interesting to observe the effect within the pressure hull of an external explosion: in slow motion records, apparently massive and rigid pieces of structure and piping flex as though they were made of rubber. Even hull valve flanges which are quite thick can be seen to distort between the stud holdings and allow water to shoot through the gaps as they flex open and shut. The studs themselves can also be stretched so that the valve appears to bounce on its seating.

The submarine design must take account of such possible motions by allowing space between equipments and the hull, so that 'collision' damage is not a contributory cause in shock failure.

FABRICATION CONSIDERATIONS

5.21   The design of a structure to withstand pressure loading should also take into account the method of fabrication of that structure. As mentioned earlier construction of the pressure hull as a cylinder makes for relatively easy fabrication because the shell plating can be rolled in a single direction to form sections of the circumference of a cylinder. A shaped pressure hull to conform to the hydrodynamic requirements can be manufactured by constructing a series of cylindrical and conical hoops which are then welded to each other around the circumference. Wherever possible circumferential welds should be associated with parts of the shell plating subjected to lower stresses, i.e. they should not be placed in a position of maximum bending stress. It is also important that there should be no conjunction of welds from different directions arriving at the same point. Thus for the plating forming each hoop, it is better if the longitudinal seams can be staggered in relative hoops so that there is no cruciform junction in the pressure hull. The welding in of inserts for penetrations should also endeavour to avoid cutting across or being close to major closure welds in the shell plating.

Even with careful temperature and atmospheric control and particular attention to sequencing, the welding of the pressure hull is liable to involve locked up stresses in the structure which must be taken into account in investigating the stress patterns within the structure. Distortion can also occur as a result of welding and this may or may not be beneficial to the structure as a pressure vessel. In this respect, there is an interesting debate about the balance of choice between the use of internal and external framing for the pressure hull.. Internal framing as generally adopted involves welding the web of the stiffeners on the inside of the shell plating. The weld shrinkage causes a slight downward movement of the shell plating between frames, giving a concertina effect. This is the character of the mode of failure discussed previously, so it may be considered that it contributes to an earlier failure of the structure. On the other hand, the welding of external frames causes the shell plating between frames to be distorted upwards slightly between frames, thereby imparting a shape to the shell plating more suited to withstand pressure. In fact, it might be argued that a series of spherical shell hoops welded together at the ring frames to form a corrugated cylinder would be a better way to design the structure, though very expensive to build. For internal framing, the welding is only really necessary to hold the frame in place and to provide some edge stiffness against buckling of the frame; the welding is not essential to the integrity of the hull as the compressive load from the shell plating would still be applied to the frame even if the weld did not exist. External frames, however, effect the load transfer between frames and shell plating by tension through the connecting weld; any defects or cracking in the weld could result in a catastrophic premature failing of the structure as a whole.

The most difficult parts of a pressure hull to manufacture are the domes at the end. For small submarines these may be a one piece plate, hot spun to form the hemispherical or torispherical shape required. However, for large pressure hulls, machines do not exist for hot spinning to the necessary size and their domes have to be fabricated from petals or segments of plating. Each petal or segment has to be pressed to the required double curvature for its position in the dome and then the pieces brought together and welded to form the total dome; the effect of the welding may be to introduce departures from the desired geometry, contributing to premature buckling.

FATIGUE

5.22   In previous sections the pressure hull structure has been regarded as an envelope withstanding a static external pressure. However, the operation of a submarine involves going from surface to maximum diving depth and various depths in between and back again. The hull is thereby subjected to an irregular cycling of the external pressure load imposed upon it. At maximum diving depth parts of the structure will experience stresses approaching the yield strength of the material. Whilst these stresses would not cause failure of the structure on a once-off basis, the operation of the vessel going a number times to or towards maximum diving depth does introduce aspects of low cycle fatigue, i.e. a small number of cycles but to high stress levels. Primary structure of the submarine will experience compressive stress at depth and this would not be expected to lead to fatigue problems, as most materials are sensitive to fatigue phenomena only in the tensile part of the cycle. But, as has been discussed, there are regions of the structure and shell plating subject to bending actions which may cause tensile stresses. Also it is possible that locked-up stresses due to fabrication and welding can lead to the material starting initially in quite a high tensile strain condition. In those circumstances the cycle from surface to maximum diving depth is not one of increasing compressive stress and strain but rather one of decreasing tensile stress and strain from a maximum at the surface to a lower value at the maximum diving depth. In consequence, it is advisable that some of the structure be considered as subject to low cycle straining on the tensile part of its fatigue curve. Mild steel is less likely to suffer from that experience because of the yielding out characteristic of the material, but present-day use of high yield, heat treated and alloyed steels which do not have that characteristic introduces the possibility that fatigue failure could occur in areas of high stress and strain cycling. It is therefore important to seek to limit the stress range encountered by every structural detail and to recognise that the structure has a finite cycle life. This may call for operational limitations in which the diving cycles in service are limited or for vessels to be reduced in diving depth in later stages of life.

CHOICE OF MATERIALS

5.23   The most common choice of material for a submarine pressure hull is steel, usually with a high tensile strength achieved by alloying or heat treatment. Steel has the advantage of a high modulus of elasticity and that makes it usually possible to design on the basis of yield stress with avoidance of buckling problems. As the strength of the steel is increased by various techniques the modulus remains constant and consequently a situation can arise in which it is no longer the yield stress condition which governs the design but the elastic buckling mode. If that were to happen it could be necessary to use more material in order to achieve the required stiffness against buckling and no advantage would be gained by the greater strength of the material. Other materials, like aluminium or reinforced plastics, can achieve high strengths but their modulus of elasticity is generally much lower and then buckling becomes the failure criterion. The fabrication of those materials also poses new difficulties. For very deep diving vessels such materials may be the only recourse, necessitating attention to the form of the shell in order to provide stiffness as well as strength. With such materials it may be necessary to consider different geometries for the pressure hull, departing from the single skin pressure envelope, for example, a double skin cellular structure, to achieve the stiffness required at depth. Greater attention has to be paid to details and penetrations, particularly where a reinforced plastic material is concerned, so that they can be carefully designed and built in at the initial stages of fabrication, as it is not a material in which it is possible to simply drill a hole and insert the penetration. However, while steel is the most commonly used material because it is relatively easy to fabricate into pressure hulls, it does have magnetic properties and as, in some instances, these can be important to the security of the vessel, the

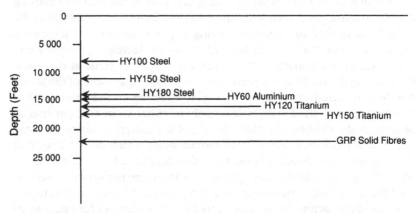

Fig. 5.14 Operating depth potential of spherical pressure hulls (weight 50% of buoyancy)

use of other materials may be determined by that consideration rather than structural arguments.

The diagram (Figure 5.14) shows the relative merits and potential of different materials with respect to achieving diving depth. This diagram indicates that for most diving depth requirements a steel hull is a satisfactory solution. It is only for very deep diving vessels that serious consideration need be given to other materials.

OTHER STRUCTURES

5.24 The emphasis in this chapter has up to now been on the pressure hull structure which constitutes the main concern in submarine design, but there are other causes for concern in other structures of the submarine.

Bulkheads

We have shown that to avoid one mode of overall collapse of the ring stiffened cylinder it is necessary to provide bulkheads at intervals along the length; usual practice is to space them at about two diameters apart. The bulkheads provide radial stiffness to the shell plating, which requires only relatively thin plating to provide sufficient in-plane stiffness relative to the shell. It should be recognised, however, that a bulkhead is subject to radial compression around its edges: how much is dependent on the relative flexural stiffness of the shell and local ring frames but it is possible for an unstiffened bulkhead plating to buckle. To avoid that, it is necessary to provide out-of-plane stiffness in the form of T-section stiffeners, so that the bulkhead becomes a plate/stiffener grillage. Though radial stiffeners would be more appropriate it is usual to use vertical stiffeners over the centre part of the bulkhead which provide supporting connections to longitudinal deck girders and suit the fitting of access doors. At the sides, intercostal horizontal stiffeners are fitted to align the stiffeners to the radial loading. (Figure 5.15)

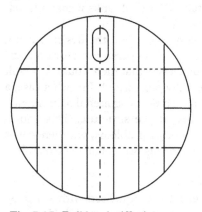 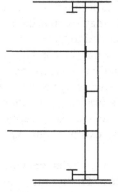

Fig. 5.15 Bulkhead stiffening

The bulkhead is an example of multiple functioning of part of the structure. It provides hull support and also deck and equipment supports. Some, but not all, of the bulkheads may also serve another function, namely those designed as escape bulkheads. The objective in that case is that, in the event of an accident, flooding within the hull will be restricted and most of the crew can retreat to the unflooded hull for possible escape or rescue. These bulkheads are therefore liable to have to withstand differential sea pressure depending on the depth of the vessel. A flat plate grillage is not ideal for withstanding pressure but it is possible to achieve pressure resistance with fairly heavy scantlings. It is also possible to adopt a different design philosophy for this once off escape situation, which hopefully never occurs. Whilst structures regularly subject to load should be designed within the elastic limits of the material, it is possible, in this case, to call upon the considerable extra capacity of the plate grillage with the material strained into the 'plastic' zone. The outcome would be permanent distortion of the structure but for a once off condition this is acceptable. It is relevant in the context of multiple functions to note that, whilst the bulkhead normally provides end support to decks and tank tops, in the escape condition the decks have to be designed to provide prop supports to the bulkhead when it is laterally loaded.

Internal tanks subject to sea pressure

5.25   Internal tanks quite often have part of their boundaries provided by the pressure hull. The ends may be the lower part of a bulkhead but the tank top will be another flat surface, giving rise to laterally loaded grillage problems similar to that of escape bulkheads. If the submarine design calls for internal tanks to be regularly subjected to sea pressure their structural design must keep strains below the elastic limit of the material, necessitating a very heavy structure. As suggested in earlier chapters that outcome is undesirable and internal tanks of that sort should be avoided. It is better to install cylindrical tanks if that alternative is acceptable, e.g. for D tanks, Q tanks.

It may be that internal tanks are incorporated into the design but not intended to be subject to sea pressure in normal operations. However, the tanks may be connected by piping in such a way that there could be a fault situation in which they become accidentally pressurised. In such a case, a 'plastic' design may be adopted, but it should be recognised that accidental pressurisation will then permanently distort the structure. This consequence may be acceptable if it caused no other problems, but there could be an exceptionally expensive repair job if the distortion had to be rectified.

External main ballast tanks

5.26   These tanks are usually arranged to be equalised with sea pressure, so a relatively light-plated structure is all that is required. However,

as they provide buoyant support to the hull when the submarine is surfaced, the structure of the external tanks and the connections to the pressure hull must be adequate to transmit the buoyancy forces. Also whilst
surfaced these tanks are exposed to waves and hence have to be designed
to withstand sea slap forces.

Though normally equalised to sea pressure, circumstances can as follows arise of a differential internal pressure in an external main ballast
tank:

(a) When floating statically, the pressure in the tank will be that of
    sea pressure at the flood holes and so the upper part of the
    tank exposed to air pressure will experience a differential of
    one or more atmospheres.

(b) During surfacing high pressure air is blown into the tank to
    displace water. For the water to flow out of the flood holes
    there must be a differential pressure (velocity proportional
    to $\sqrt{\Delta p}$). The volumetric rate of water discharge must equal
    the rate of volumetric expansion of the air. Thus a differential
    pressure will be necessarily created within the tank, the mag-
    nitude of which will depend on the rate and pressure of the air
    discharge and the area of the flood holes. This pressure has to
    be catered for in the structural design of the tank.

(c) If MBTs are blown deep as part of an emergency recovery the
    air being admitted to the tanks may not fully empty them of
    water. As the boat, hopefully, rises, this air will expand as sea
    pressure reduces but the expansion will be limited as in (b) by
    the rate of discharge of water. If the boat accelerates and
    surfaces rapidly the water will not have discharged in time and
    there will be a significant over-pressure in the tanks. To allow
    for this is a complex calculation that would be performed in
    detailed design, but may lead to a heavier structure than the
    other conditions might lead the designer to expect.

(d) External fuel tanks would normally be equalised to sea
    pressure at all times and not be subject to the internal pressure
    differentials of MBTs. They have to withstand the more
    common loads experienced by the external hull and so may
    well be of similar scantlings to MBTs. They are sometimes
    fitted with internal partial bulkheads to form a labyrinth
    between the fuel intake and the water inlet, which reduces
    sloshing and mixing of the fuel/water by reducing the area of
    the interface.

Other hull structure

5.27   The fore end free flood structure is also equalised with sea pressure but may take the brunt of wave loading. Internally this structure

tends to be complicated as it partly forms MBTs, while part may be Torpedo Discharge Tanks which are also subject to internal over-pressure. The structure also provides support to the bow sonar, torpedo tubes, forward hydroplanes and anchoring arrangements. Some of these represent quite heavy loads on the fore end structure, acting as a cantilever support from the main hull.

One of the more difficult structural connections is that with the exposed pressure hull. Here the fore end structure has to scarph in to the hull to provide a hydrodynamically smooth transition. As this occurs near the dome/cone connection of the hull an acute angle junction is created, liable to cause severe welding problems. One solution is to provide a forged connecting ring between dome and cone with an extended lip to pick up the fore end plating. (Figure 5.16)

Similar problems occur with the after end structure, which may partly form MBTs and also have to provide support and transmit the control forces from rudders and hydroplanes, tail shaft bearing and the propulsor. With the narrower cone angle of the tail the hull/after end structure connection may be even more acute than at the fore end.

Superstructure

5.28   Though much less than for a surface ship, a submarine usually requires some superstructure. This takes the form of a casing to house piping, cables, deck arrangements, hull hatches, etc. in a streamlined containment, and also to provide a walk way for crew in harbour. In double hull or reduced diameter hull sections this casing may be part of the external circular section. In other forms the casing is at a height above the pressure hull. It is important to save weight at that height and though steel casings have been common, aluminium has also been used and

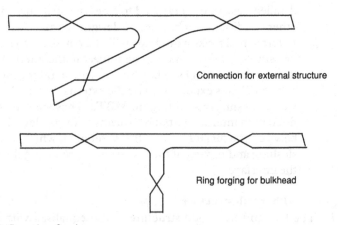

Connection for external structure

Ring forging for bulkhead

Fig. 5.16 Junction forgings

more frequently nowadays glass reinforced plastic, GRP, is used. Apart from being relatively light and strong it does not suffer from corrosion problems which are more serious at the water/air interface. The primary strength criterion is again that of sea loading, as there should be sufficient venting and flooding holes to avoid differential pressure.

The bridge fin provides top bearing support to masts and periscopes, houses the high conning tower access hatch and provides a surface bridge conning position. It may also have to support fin mounted hydroplanes if this configuration is adopted, and there may be other sensors and communications aerials housed within the fin. As we shall see in the ensuing chapters, the bridge fin is highly undesirable hydrodynamically but as yet no other solution has been found for housing the many demands for top side equipment. It is usual to provide strong steel supports to the masts, etc. The sheathing of the fin may be of steel, aluminium or GRP. Some pressure loading can arise if the submarine surfaces rapidly, causing a temporary head of water within the fin.

One reason for providing an all steel superstructure is the operation of surfacing through ice, when the buoyancy force of the hull is transmitted through the fin and later the casing to break what is intended to be thin ice. A limiting load condition for design can be estimated from the vertical force on the top of the fin which would cause unstable heeling of the vessel, so that this force is related to the size and transverse stability of the submarine.

SUMMARY OF STRUCTURAL DESIGN PHILOSOPHY

5.29   The overriding philosophy in pressure hull design is, we suggest, to retain the simplicity of the pressure-carrying envelope, avoiding the possibility of high stresses due to complex strain geometries, and at the same time providing a geometry which affords adequate stiffness in each of the several modes of buckling failure.

The details of structural failure discussed earlier are not directly of import in initial design. Nevertheless, the submarine designer must be aware of these factors, which impose constraints of either an essential or a desirable nature in the choice of hull configuration adopted.

Thus in this chapter there is strong advice to the designer to avoid changes of pressure diameter along the length of the hull. There may, however, be dominant overall design considerations which lead to the decision to adopt a configuration incorporating step changes in diameter. This decision must be made in the knowledge of the associated structural complications and fatigue implications which follow. Examples are given in the book where that decision has clearly been made by the designers, so palpably it is possible to adopt this configuration and solve the structural problems. The questions to be answered in any new design are, is it necessary and does the benefit justify the cost in general terms?

5.30 We have avoided entering into a detailed account of the state of the art in structural analysis for submarine pressure hulls, because it is a specialised field in which there have been in the last few decades considerable advances, particularly through computer-based numerical analysis methods. However, despite the precision of these methods, the uncertainties due to the differences between real pressure hulls and the analytical models are still sufficient to warrant proceeding conservatively. That approach entails treating failure modes which involve a greater degree of uncertainty with more caution than those modes with a lower degree of uncertainty. That line of thought encourages the idea of ranking the main failure modes in order of greater to lesser uncertainty for which we suggest the following:

    Collapse of dome bulkheads
    Overall collapse of long cylinders
    Tripping of ring frames
    Interframe collapse

5.31 With a factor of safety approach to structural design, the response to that sort of ranking is to apply the largest factors to the modes with the greatest uncertainty. (There is an alternative, probalistic, approach but it is less helpful to our present purpose). That philosophy leads in the design of pressure hull structures to aiming to ensure that the hull will fail by interframe collapse. The emphasis then falls on enabling the designer to evaluate the interframe collapse pressure of a pressure hull of given geometry as accurately as possible. That issue is discussed further in Appendix 4.

5.32 As we have observed, there are now readily available computer programs using the scantlings of a pressure hull of given geometry and material which can be systematically investigated. The search would be for an optimum design of the structure, including on assessment of the sensitivity of the findings to dimensional variation so that the submarine designer can be aware of the penalties in weight terms if, for other than structural reasons, e.g. convenience in layout or in fabrication, he were to depart from optimum scantlings. Significantly in that process, catering for adequate margins of structural safety in relation to the various failure modes serves to provide the following information for given overall geometry and material of the pressure hull:

    Interframe collapse: shell plating thickness as a function of diameter and frame spacing.

Overall collapse: frame area and bulkhead spacing as a function of diameter.

Frame tripping: frame proportions and moments of inertia.

Dome bulkhead collapse: dome plating thickness as a function of curvature.

In the previous chapter it was suggested that initial design is based on space determinants and that weight may be manipulated by a suitable choice of diving depth to govern the dominant weight component of the pressure hull.

In the foregoing discussion emphasis has been placed on the modes of failure which may occur, and which are taken into account in the detailed structural design to provide a predicted collapse depth of the hull. At the beginning of the chapter discussion centred on the requirements for operating depth. These two considerations are related by an overall factor of safety for which the concept designer must seek guidance from the standards, which have been set after extensive investigations. The selection of an appropriate safety factor involves consideration of a number of aspects. There is the engineering consideration of the accuracy with which the collapse depth can be predicted; then there is the probability of the submarine accidentally exceeding its operating depth due to control failure and how large is the likely excursion; there is also the question of how often the submarine will approach maximum operating depth and the effect on fatigue life of the hull structure.

These considerations may lead to the definition of a depth to which the submarine may operate freely and a deeper depth to which the submarine may go under very strictly controlled conditions.

6 POWERING OF SUBMARINES

INTRODUCTION

6.1 The powering of a submarine vehicle is one of the most important factors in the determination of its size. As shown in Chapter 4, power plants use a high proportion of the weight and space available in a submarine, some 35% of weight and 50% of the total volume being devoted to power generation and storage. As we go on to show, the power requirements are determined in conjunction with speed by the size of the vessel, and hence the designer encounters a loop in the design process whereby the output of the power assessment in terms of a volume requirement for propulsion plant is itself a significant input to that assessment. In control engineering terms this is a positive feed-back loop, which can all too easily cause a growth in the total size of the design.

The powering of a submerged conventional submarine is an exercise in efficient energy storage and its conversion to usable power. It is the solution of this problem which dictates the underwater endurance of the submarine. As described earlier, it was only when a nuclear power plant became available that a true submarine could be produced, because it was the development of an energy source totally independent of atmosphere that enabled the true submarine to become a reality. The successful production of a practical nuclear power source not only enabled the design of a true submarine but also shifted the endurance limits to other factors such as the crew and expendable stores, because the nuclear power source endurance is measured in years rather than days or weeks. However, there is a price to be paid in that the vessel must inherently be large to accommodate the reactor and all its safety features. Reactor plants also tend to come in specific sized units which tie the design to one of the available sizes. Thus nuclear powered submarines tend to start large and so the application is to high performance, high capability designs where there is a demand for large energy consumption.

For smaller, low cost, submarines nuclear power in its present form is not an option and even for the larger nuclear powered vessels there remains a requirement to exercise economy in the way the design consumes power. Consequently, we are able in this chapter to discuss the design considerations bearing on power consumption almost independently of the choice of power plant. Since, however, choice has to be made as a result of the powering assessment, in the latter part of the chapter we examine the impact on design of different power plant and energy storage systems.

STATEMENT OF REQUIREMENTS

6.2 As with every other design feature the powering assessment stems from the users' requirements. Power and energy storage are almost entirely dependent on the vehicular performance requirements of the submarine. Though the operational equipments place some demands for energy provision and must be included in what is known as the Hotel Load, i.e. that required to keep the submarine 'alive', their demand is usually quite a small proportion of the total energy requirements. However, to extend the time fully submerged the propulsion demand may have to be reduced to a minimum and then it is the Hotel Load drain on the energy storage which will determine the actual time submerged.

The primary factors which govern the power and energy storage requirements are: the maximum speed and for how long; the range of operations, i.e. distance from base to operational area: and the submerged endurance, i.e. how long the submarine is required to remain isolated from the surface.

6.3 The maximum speed of a submarine can be the most difficult and contentious aspect of the dialogue between operator and designer. This is because it is difficult to find a logical reasoning to arrive at a required maximum speed. Yet it is probably the most expensive requirement to meet in a design. The propulsive power requirements for a submerged submarine of given displacement vary as the cube of speed. In essence, to double the maximum speed of a submarine from 20kts to 40kts would require eight times the propulsive power; in fact, as the submarine would have to be considerably larger to accommodate an eightfold increase in power plant, the speed achieved would fall short of the 40 kts and so even more power and size would be called for. The designer could be placed in a position of having to oppose unsubstantiated requirements for high maximum speed. From a military operations viewpoint having high speed available is always desirable, because with a speed advantage the submarine may be able to outrun its attackers or, if attacking, close rapidly on its target. Since submarine warfare tactics are largely involved with stealth it can be argued that use of high speed in evasion or attack is not the only, or necessarily the optimum, tactic. The risk of an operator/designer impasse could be avoided by the designer adopting a pragmatic approach and attaining as high a speed as possible with the limited available choice of power plants and energy storage devices. A more logical approach would be to try to establish the minimum/maximum speed capability which would permit the submarine to perform its required functions. To arrive at such a figure requires the replacement of a velocity parameter, speed, by the related parameters of distance and time. Thus in the scenarios in which the submarine is required to operate, circumstances can arise in which there is a requirement to move from position A to position B and a time factor to make such a move worthwhile. For example, the submarine may be required to transit from base to its operational area and to achieve passage within a given time. Thus a mean speed of transit is determined and, allowing for times when the submarine has to slow down, a minimum transit speed is identified. Alternatively, a submarine in its operational area may need to move to an intercept position in order to re-acquire or attack a target within a given time. This could lead to another min/max speed requirement as discussed in Chapter 11.

Requirements expressed in that way are more useful in the design context because they allow trade-off considerations. The possible cost consequences of the speeds being considered can be weighed against the consequences of non-achievement, e.g. additional submarines may be required to give cover in the operational area if relieving submarines take too long to get there.

6.4   Whilst maximum speed requirements determine the size of the propulsion plant needed, it is the time at speed which can govern the energy storage capacity requirements of a submarine. Because with a nuclear power plant the energy is effectively unlimited in terms of a patrol period, it is quite practical for a nuclear submarine to transit for thousands of miles at high speed. On the other hand, a battery powered submarine may only be able to run at maximum speed for half an hour before the battery is exhausted, when it would have to surface or snort on diesel engines to recharge the batteries, and then it is important to establish the requirements for time associated with the speed pattern. For non-nuclear powered submarines the range of operations determines the quantity of fuel to be carried at the start of a patrol. A submarine will have to spend some of the transit time at snorting depth so as to be able to run in an air breathing mode. During those periods the power output of the diesel generators will have to provide the power for propulsion, the Hotel Load and the power to recharge the batteries for subsequent submerged running. That necessity results in periods when the vessel is vulnerable to detection from the surface or air surveillance, which leads to an operational requirement to identify the 'indiscretion ratio' of the design, i.e. the proportion of transit time when the vessel is snorting.

6.5   The other requirement to be determined is the submerged endurance of the vessel, i.e. how long can it remain fully submerged, which depends upon the energy storage capacity and the speeds required whilst submerged. If the submarine is not attempting to move then the drain on energy storage is just that to meet the Hotel Load, which includes any energy consumption needed to sustain a breathable atmosphere. Even a conventional battery submarine can remain submerged for a long time if stopped, whilst there is effectively no limit with nuclear power. However, if the vessel has to sustain speed submerged then, as already observed, the energy consumption will vary as the cube of that speed; the higher the speed, the less time the vessel is able to stay down before needing to replenish its energy storage.

6.6   Other underwater powering states to be considered are:

(a) Listening speed

The submarine's main sensor underwater is its sonar. Even if in a non-action state, it needs to know whether there are other vessels in the vicinity, possibly on collision course, or whether topographical features of the sea bed or under ice may be in its path. At high speed the self and flow noise generated by the submarine may effectively deafen the vessel's own acoustic sensors. Thus it has either to transit at slower speed or slow up at regular intervals to check the situation. It is then highly desirable that the design should take into consideration maximising the speed at which acoustic sensors remain effective, which entails reducing the transmission of on-board noise makers and reducing the fluid flow noise, particularly over the parts of the hull where the sonar transducers are located.

(b) Quiet speed

Here the submarine aims to minimise the range at which it might be detected. Apart from any measures taken to reduce the radiation of noise, this state usually necessitates a reduction of power demand so that many of the potential sources of noise, e.g. both propulsive and auxiliary machinery, can be shut down or run at low speed.

(c) Ultra quiet

In this state the submarine takes every possible step to avoid acoustic detection and to maximise its capability to detect other craft, which usually entails reducing speed to the minimum consistent with maintaining control and shutting down virtually all operating machinery on board, including ventilation and air-conditioning. This state cannot be sustained for very long, but it is important that the design of all systems should be consistent to enable it to be accomplished. For example, there would be little point in shutting down the ventilation system if the hydraulic system needed frequent operation of the pumps. The state also necessitates that special noise reduction measures should be taken on those items of machinery and systems that are essential to operate continuously or at regular intervals.

RESISTANCE TO MOTION

6.7 In summary, there are four desirable features which a submarine should have in order to be able to meet the foregoing requirements to best advantage:

A low hydrodynamic drag of the hull

A high efficiency of the propulsive system

A high energy conversion efficiency of the power plant

A high energy density storage system.

We now go on to consider those factors which affect the resistance to motion of the submarine when underway.

Submerged drag

In general there will be little conflict between the hydrodynamic requirements appropriate to the various speed regimes considered in the earlier section. A minimum drag body for high speed will be compatible with that for quiet operations at slower speeds.

In the deeply submerged condition all drag is associated with the viscosity, albeit low, of the water. It is customary to regard the drag of the bare hull as having two components, described below.

(a) Skin friction drag

The viscous shear drag of water flowing tangentially over the surface of the hull contributes to the resistance of the hull. Essentially this is related to the exposed surface area and the velocities over the hull. Hence, in general terms, for a given volume of hull it is desirable to reduce the surface area as much as possible. However, it is also important to retain a smooth surface, to avoid roughnesses and sharp discontinuities and to have a slowly varying form so that no adverse pressure gradients are built up causing increased drag through separation of the flow from the hull.

(b) Form drag

A second effect of the viscous action on the hull is to reduce the pressure recovery associated with non-viscous flow over a body in motion. In an ideal non-viscous flow there is no resistance since, although there are pressure differences between the bow and stern, the net result is a zero force in the direction of motion. Due to the action of viscosity there is reduction in the momentum of the fluid and, whereas there is a pressure build up over the bow of the submarine, the corresponding pressure recovery at the stern is reduced resulting in a net resistance in the direction of motion. This form drag can be minimised by having very slowly varying sections over a long body, i.e. tending towards a needle shaped body even though it would have a high surface to volume ratio.

It can be seen that because of the viscosity of the water these two components lead to opposing requirements for the shape of the body for least resistance. For a given volume of displacement, as the body becomes longer and thinner, the form drag is reduced, but as the surface area increases the friction drag is increased. (Figure 6.1) In consequence, there is an optimum, albeit very flat, where the length to diameter ratio of the form is approximately 6:1, at which the total drag is a minimum.

An ideal underwater shape for low drag is a cigar shaped form, (Figure 6.2(a)) circular in cross-section and usually with an elliptic bow and a parabolic stern coming to a point as shown. This is characteristic of the shape of airships earlier in the century.

6.8   Although it might seem that this form is the ideal at which all submarine designs should be aimed, in practice there are many considerations in overall design which can lead the designer to depart from the ideal. It could only be in a vehicle, like a test vehicle, solely aimed at minimising resistance, that such a form could be adopted above all other considerations. In an overall form it may be that volume distribution considerations inside the hull lead the designer to depart from the ideal

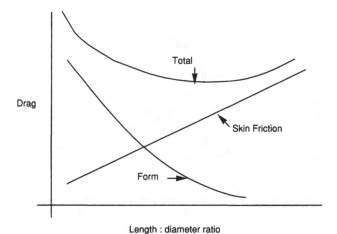

Fig. 6.1   Drag components for constant volume form

length/diameter ratio. Given a specific total volume, the ideal might call for a larger diameter than could be usefully deployed, and also a shorter length than needed to meet the demands for equipment length, such as propulsion plant and weapons stowage. Thus it is quite common to find that submarines designed for the submerged mode do have a length to diameter ratio appreciably larger than the optimum. (Figure 6.3) As demonstrated by Figure 6.1, the total drag curve is very flat near its minimum and so little penalty is actually entailed in modest departure from the ideal. The ideal form also involves a continuously changing diameter along the whole of the length which adds to high production costs of the hull. To avoid that penalty it is usual to introduce an element of parallel mid-body into the form. (Figure 6.2(b)) It transpires that this departure

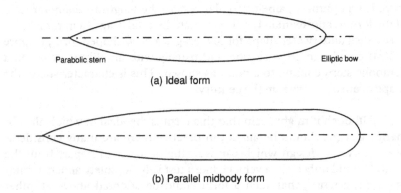

(a) Ideal form

(b) Parallel midbody form

Fig. 6.2   Hull forms

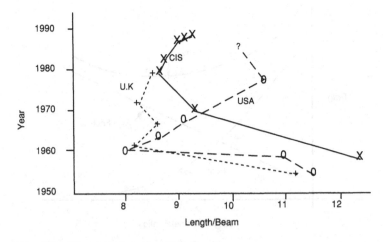

Fig. 6.3   Trends in length to beam ratio

from optimum does not incur a heavy penalty. A continuously curved shape would also require a tailor-made cradle for docking the vessel. The alternative would be to provide a parallel flat-bottomed keel, but that would add to the drag as an appendage to the hull. Another alternative arrangement would be to have a parallel pressure hull enveloped in a thin skinned outer envelope shaped to suit the ideal form, but adoption of that configuration would involve considerable extra volume external to the pressure hull; unless there were good justifications in terms of high reserve of buoyancy in ballast tanks and large fuel capacity, such a configuration would increase the total volume to be propelled and may increase the total power required despite the better resistance form.

The ideal shape may also have to be departed from in detail. Though it is normally recommended that the bow should be of an elliptical form, consideration has also to be given to siting the sonar array and making provision for weapon storage/discharge, forward hydroplanes or such mundane aspects as the anchoring arrangements. These may lead the designer well away from the optimum axisymmetric elliptical bow form. The alternative could be to house all such items within a nicely shaped bow, but this would again be at the expense of adding to the external form volume.

The slowly tapering curved stern shape recommended can also give problems in a real design. It involves a structural problem in connecting the tail cone external to the pressure hull at a shallow angle, and also results in low volume aft where machinery and propulsion systems usually call for large space. This can lead to the acceptance of a fuller form aft than would otherwise be considered ideal. However, there is a trade-off in, perhaps, a poorer form against avoidance of the larger form needed to get the shape 'right'. Also, as we shall see later, there is another trade-off with propulsor efficiency, which can be improved by the fuller stern sections.

Control of submarines is considered in detail in Chapter 8, but there is a design interaction we need to address in this section. The control requirement is for directional stability; the basic shape so far considered is, on its own, directionally unstable, with the shorter fatter form least stable. To provide stability, fins are fitted at the stern, which increase the drag. It may be that a longer slender body will entail less drag from its required stabilizers. Hence it is not necessarily true that the 'ideal form' gives the least resistance form when stabilisation is taken into account.

6.9   In any case, a submarine is unlikely to have a simple axisymmetric form externally. Requirements exist for control surfaces, a bridge fin, very often a casing, sonar extensions plus a myriad of inlets and outlets for taking in or expelling water. Even when these are faired into the hull they lead to additional drag items, which very often are disproportionately high relative to their size on the main body. (Table 6.1)

Table 6.1 *Components of resistance*

Component	$\dfrac{\Delta P_{\rm E}}{\text{Total } P_{\rm E}}(\%) = A$	$\dfrac{\Delta \text{ Area}}{\text{Total Area}}(\%) = B$	Resistfulness $\dfrac{A}{B}$
Hull	68.50	84.33	0.81
Bridge fin	7.87	8.05	0.98
Stern planes	7.71	3.28	2.35
Bow planes	3.54	0.58	6.14
Upper rudder	5.34	1.64	3.26
Lower rudder	1.81	1.11	1.63
Sonar fairing	2.88	0.16	18.03
Ballast keel	6.05	6.10	0.99
Total	103.70		

Older conventional submarines, which gave less emphasis to submerged high speed performance, typically had a total drag of more than twice that of their bare hull form due to the non-streamlined appendages and, in particular, the holes provided for free drainage of casing and bridge fin and flood holes in the MBTs. Even with modern 'streamlined' boats there can be an additional 20% or more greater drag over the bare hull drag due to appendages. It behoves the designer from the hydrodynamic aspect to reduce these parasitic drags due to departures from the simple bare hull form. This need involves much attention to the fairing of the appendages in their own right and particularly their fairing into the hull to avoid separation drag at the root of the connection.

Though the operational necessity to dive and surface rapidly has diminished with the modern submarine operating primarily submerged, there is still a requirement for rapid drain down of free flood volumes at the top of the hull to avoid stability problems and to allow water exit from and ingress into the MBTs. Both involve providing quite large openings in the otherwise smooth hull of the vessel. Special attention has to be paid to their shaping and alignment to avoid additional drag. The main cause of drag due to the flow across a hole is the induced fluctuation of flow in and out of the hole caused by pressure differences and instabilities, which is exacerbated by the forward-looking edge at the aft end of the hole acting as a splitter plate and causing fluctuating eddies to be shed. It is a similar process to that used deliberately to create sound in a flute or organ pipe. The phenomenon can happen in a submarine at speed if the connecting tank or internal volume experiences a resonant condition with the eddy shedding. Not only is energy lost from the flow contributing drag but noise may be radiated as well. In some extreme instances the resonance of the structure has resulted in fatigue cracking. To overcome those effects closure doors have been fitted to major openings, but they have to be articulated to allow the ingress and egress of water as required at the opening. Because the mechanism is invariably permanently submerged, it involves constant maintenance effort. A simpler solution is to fit a flat bar grill across the opening with the bars across the flow direction and angled away from the flow. These effectively inhibit the oscillatory eddy shedding whilst imposing little restriction on the necessary flow of water through the hole

It is important that as far as possible any deviations from the ideal form in the way of appendages or lumps should be incorporated within a smooth overall envelope rather than as obvious additions, but this can lead to added expense in construction and increases in form displacement, though overall it is likely to result in reduced drag.

6.10   In general then the designer should aim for an overall hydrodynamically smooth form, albeit with appendages, even though this may be a considerable departure from the optimised or ideal form described at the beginning of this section. Although the main emphasis in the section has been on the reduction of resistance drag on the hull, there are other factors to be considered which are consistent with the same philosophy. One is the reduction of turbulent flow over acoustic sensors as this is liable to blanket and so reduce their effectiveness; the other is that the flow noise may be sufficient to cause a radiated signal liable to detection. The flow over the hull inevitably causes variations in the flow into the propulsor, which is both turbulent and variable in velocity, and which can lead to a noisier propulsor. We return to these propulsor effects in a later section.

There is an implicit assumption in the foregoing that the submarine moves in a direction along its axis. On occasions, however, because of hydrostatic imbalance, the submarine may need to proceed at an angle of attack. When submarines travel along such an off-condition path, their resistance will be augmented by crossflow drag components.

SPEED–POWER RELATIONSHIP

6.11   In general terms the resistance of a body to motion through a fluid can be expressed in the formulation $R = C_D \rho_f A U^2$, where ρ_f is the density of the fluid, A is a representative area of the body, U is the speed and C_D is a Drag Coefficient related to the shape of the body. In many applications the representative area is the cross-section or presented area of the body to the direction of motion and values of C_D are quoted on this basis. For marine vehicles, however, it is more common to adopt the representative area as the surface area exposed to the flow, with appropriate values of C_D. In early stages of design this surface area has not been determined and a more useful form of the formula is related to the volume of the body. Taking the cube root of the volume provides a representative length and squaring this value provides another measure of area. Hence for early estimates it is possible to represent the resistance of a submerged submarine in the form $R = K \times Vol^{\frac{2}{3}} \times U^2$ where Vol is the Form Volume and K is a coefficient which depends on the shape and configuration of the hull as just discussed. Suitable value of K can be derived from model tests on previous submarine forms.

Since the power is given by Force × Distance/Time the power required to propel a submarine can be expressed as $P_E = K \times Vol^{\frac{2}{3}} \times U^3$.

Thus for a given characteristic shape the powering of a submarine depends on the size of the form volume and most significantly on the cube of the speed required.

SURFACE RESISTANCE

6.12   Although the modern submarine is designed with most emphasis on submerged performance it nevertheless has to operate on the surface on occasions, such as entering and leaving harbour, and sometimes for lengthy transit passages from base to its diving area. Problems of resistance whilst running on the surface are similar to those of normal surface ships except that for a submarine the majority of the displacement is below the water with very little freeboard above the water. Most submarines are also relatively small vessels and this means that to make any reasonable speed on the surface they are operating at a high Froude Number, so that the wave making component becomes dominant in surface resistance.

The short fat form idealised for underwater performance is unsuitable

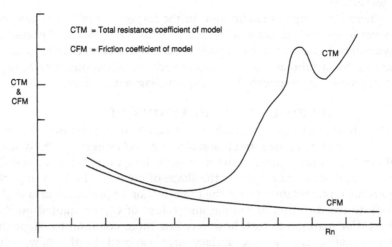

Fig. 6.4   Typical submarine resistance coefficient curve – surface condition

for this speed range as it leads to operation near the main hump of the resistance curve. An increase in length would be required to bring the vessel back down the steep part of the resistance curve. Many submarines in the first part of this century were, effectively, submersibles and conducted most of their transit on the surface. Therefore they typically had more of a ship shaped hull; longer and thinner and with a pointed bow sometimes incorporating flare above the water. With this form an older conventional submarine was able to move faster on the surface than some modern, more powerful submarines operating out of their submerged design environment. (Figure 6.5) The fact that other design considerations may lead the designer of a submerged vessel to adopt a longer thinner form than ideal, with the penalty submerged being not too great, gives some advantage to that vessel when running on the surface; but perhaps not as much as could be achieved by deliberately designing for surface running. It is apparent that a flared and pointed bow is not a good shape for underwater operations, but on the other hand the full form elliptic bow optimised for underwater performance has the undesirable characteristic on the surface in that it causes a large upwelling of the bow wave across the foredeck, sometimes reaching as far back as the bridge fin. Associated with this is a tendency for the hull to be driven under the water, which requires the boat either to run at a large stern trim angle or even to use its forward hydroplanes as a means of keeping the bow up in the water; both these effects add to the surface resistance.

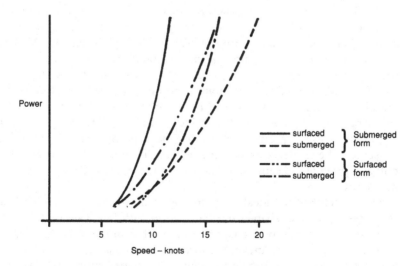

Fig. 6.5   Relative speed/power curves – surfaced and submerged

As we have observed, virtually all modern submarines are optimised for their underwater performance, and so it is usually accepted that surfaced performance has to be compromised. However, in some situations the operational requirements may render such a compromise unacceptable, in which case the designer has to seek a balance between submerged and surface performance.

6.13   A conventionally powered boat has to spend a proportion of time on patrol snorting and therefore has to run in a sub-surface condition at snort or periscope depth, which is fairly close to the surface. In this state it is affected by waves at the surface which will add to the drag by causing lateral motions of the hull and there will also be additional drag due to masts and periscopes caused by both the parts below the water and by the parts at the surface making waves and plumes. Though the main hull itself is submerged it is still near enough to the surface to have some reduced wavemaking effect on the surface; the wavemaking pattern is not as pronounced as that of a vessel running on the surface, but there will nevertheless be a distinct wave pattern and wave interference effect similar to those evinced by a surface vessel, which further adds to the resistance of the submarine in the snort condition. The effect is important because at this stage the boat is not only having to propel itself but also to provide sufficient additional power to charge the batteries and service the Hotel Load. In consequence, snorting is the definitive power condition for generators which has to be considered in the design.

6.14   To establish the satisfactoriness of the hull form a thorough hydrodynamic investigation has to be carried out. The usual method, still applied today, is the testing of scale models pioneered by William Froude. By towing scale models the resistance can be measured and using appropriate scaling laws the resistance of the full scale vessel can be predicted. This bald statement hides a number of complex questions on how scaling should be accomplished as, particularly for submerged bodies, it is difficult to achieve similarity. The designer has to be guided by the expertise of the experimenters. Apart from predicting resistance, the model tests also enable visualisation of the flow over the hull and provide a means, albeit *ad hoc*, of reshaping the body to obtain better flow.

The precepts of the ideal form are the result of extensive methodical series testing of hull shapes and correlation with full scale trials, as we indicated in the historical chapter. Many other model tests have since been conducted on variants of the form, so there is a large data base and wealth of experience for the essentially cigar shaped submarine. As previously emphasised in the structural chapter, the designer may have reasons to make major departures from this form, but in doing so he would require extensive research and testing to obtain sufficient data to enable reliable estimates to be made.

An alternative approach to physical model testing is by means of Computational Fluid Dynamics modelling, usually termed CFD. It is becoming possible with powerful computers to determine the flow around a defined but arbitrary shaped body in motion. This is a complex mathematical subject and results obtained have not yet all been fully validated. However, the approach holds the hope that in the future the designer may be able rapidly to assess flow characteristics of his proposed hull form and incorporate beneficial changes at an earlier stage of design.

Whichever method is used, it is necessary first to postulate the shape and then to assess its performance. If there are deficiencies in that assessment neither method provides much indication of how to correct for them. It is then necessary to modify by intuition or experience and try again until a satisfactory solution is found.

It is for consideration in the future whether the power of computers would enable a different approach to be developed, which would be to use a direct hydrodynamic shaping of the hull to achieve what might be called true streamlining.

PROPULSION

Propulsive efficiency

6.15   The design of the propulsor for a submarine is a specialist task for a later stage of design. At the early stages of sizing and shaping the vessel the important factor is the expected propulsive efficiency which with the estimated resistance determines the power required and hence the size of the propulsion plant. The propulsive efficiency has essentially three parts in the traditional method of approach.

The first and major part is the efficiency of the device itself. The purpose of the propulsor is to develop thrust to overcome the resistance to motion of the vessel, so the delivered power of the propulsor is given by thrust times speed. It will be assumed initially that we are discussing a screw propeller, the thrust of which is developed by blading rotating about a shaft. To generate the thrust, energy must be input to the shaft from the propulsion motor. This shaft energy at the propeller is the input energy and so we can describe its efficiency as the ouput (thrust) energy divided by the input (shaft) energy.

An alternative more general method of assessing efficiency is to consider the action of the propulsor on the fluid. To develop thrust, fluid is accelerated through the propulsor causing a change of momentum of the fluid, which can be assessed as the mass flow through the propulsor times the change in velocity of that mass of fluid from upstream to down stream, for example:

$$T = \rho A_p V_p \times (V_D - V_U)$$

The energy expended in producing this thrust is the change of energy of the fluid from upstream to downstream, for example:

$$P = \frac{1}{2}\rho A_p V_p \times \left(V_D^2 - V_U^2\right)$$

An immediate corollary of these expressions is that to develop a given thrust, less energy is expended if a large mass of water (area of propulsor) is given a small change of velocity. Hence for propulsor efficiency there is benefit (taken in isolation) in having a large diameter propeller.

6.16   The second part of the overall propulsive efficiency arises from the interaction between the propulsor and the hull which is in close proximity.

The foregoing discussion on propulsive efficiency treated it in isolation. When the propulsor is close to the hull three separate effects arise, as follows:

(a)   If the speed of advance of the vessel is V this might be taken to be the upstream velocity of fluid approaching the propulsor V_U. However, as the fluid passes around the hull it changes velocity. At the bow it comes virtually to a standstill at the stagnation point. It then accelerates around the bow and has a speed slightly higher than the speed of advance around the sides. As the stern form reduces in diameter the fluid again slows down. In addition to the shape effects of the body on fluid velocity the viscosity drag effects on the hull surface also cause a slowing of fluid velocity. Hence at the stern there is an area surrounding the tail of the hull where there is slow moving water, which is termed the wake. The wake fraction is defined as

$$w_T = \frac{V - V_u}{V}$$

If the propulsor is sited in the wake the upstream velocity V_U may be considerably less than the speed of advance. A second corollary of the simple analysis of propulsor efficiency is that less energy is expended if the absolute or mean velocity on which a change is brought about is low. Hence there is an efficiency advantage in siting the propulsor in the low velocity wake at the stern of the vessel.

It was concluded above that a large diameter propeller can be expected to give higher efficiency, and we can now add something for the wake effect. However, the wake is limited to a region around the stern, so a very large propeller diameter may extend beyond the wake and lose some of the effect resulting in rather poorer overall efficiency. Hence there is a limit to the hydrodynamic gain with a very large diameter propeller. (Figures 6.6 and 6.7)

(b) The propulsor develops its thrust by accelerating fluid through it. This acceleration effect can be shown to extend forward for one to two diameters. Therefore the stern of the vessel experiences higher velocity flow over the hull than if towed. This causes an increase in drag known as the augment of resistance due to the propulsor. The propulsor therefore has to provide more thrust T than the bare resistance of the

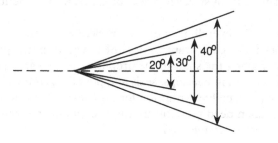

Fig. 6.6   Tailcone angle

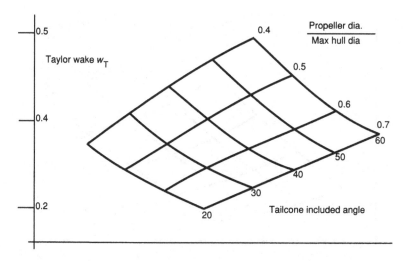

Fig. 6.7   Effects of tailcone angle on wake fraction

hull R. Though best treated as an increase in resistance it is more common for the effect to be treated as a loss of effective thrust, i.e. as a Thrust Deduction, $t = T–R/T$ which can be understood in terms that in developing thrust the propulsor offsets some of its thrust by action on the hull. By moving the propulsor further away from the stern the associated loss of efficiency can be reduced, but to do so may lose the advantage of the wake. (Figure 6.8)

(c) The third interaction effect also concerns the wake. For a hull which is not exactly axisymmetric and on which there are a number of appendages, the wake is not uniform. As well as varying in velocity radially there will be circumferential variations due to wake 'shadows' from upstream appendages. The rotating blades of a propeller thus encounter changes in velocity with a resulting loss of efficiency. Though the effect should be small, it must nevertheless be taken into account. (Figure 6.9)

Part of overall propulsive efficiency are the mechanical losses incurred in transmitting the power from the propulsion motor along the shaft to the propulsor. These are mainly frictional losses in bearings, seals and possibly gearing if installed. Though usually accounting for only a small loss in efficiency they again need to be taken into account in the configuration of the stern arrangements.

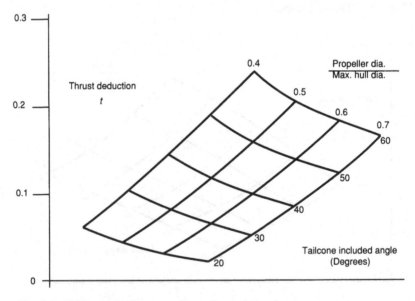

Fig. 6.8   Effects of tailcone angle on thrust deduction

Table 6.2 *Effect of tailcone included angle on PC and P_{Shaft}*

		w_T	t	η_H	η_o	η_R	PC
Single Screw	20° L	0.26	0.04	1.30	0.65	1.02	0.86
	40° L	0.36	0.11	1.41	0.65	1.02	0.93
	60° L	0.46	0.18	1.52	0.65	1.02	1.00
Twin-Screw	—		0.20	0.15	1.06	0.65	0.99	0.68

$$\eta_H = \frac{1 - t}{1 - w_T}$$

$PC = \eta_H \eta_o \eta_R$ = Propulsive efficiency coefficient

Single open screw propeller

6.17   The foregoing brief and simplified explanation of the complex problem of propulsor design presents the background the submarine designer needs in deciding on the propulsor configuration. It serves to indicate the advantages in propulsive efficiency terms of the single axial

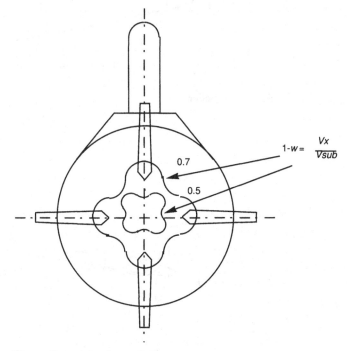

Fig. 6.9   Circumferential wake variation

screw propeller of fairly large diameter, and gives some reasons for its adoption in the *Albacore* experimental high speed submarine and subsequently in many modern submarines. With that configuration overall propulsive efficiencies (PC) of 70-80% can be achieved in submarines compared to about 60% in twin screw surface ships. (Table 6.2) The latter figure would be expected in an older twin screw submarine of World War II vintage, the configuration of which loses out both by having small diameter (limited by clearance from the hull) and enjoying less advantage from the wake, whilst suffering more from non-uniform wake distribution, because when mounted below or to the sides of the hull the propeller blades encounter the wake on the inboard side of the disc and open full velocity flow on the outboard side.

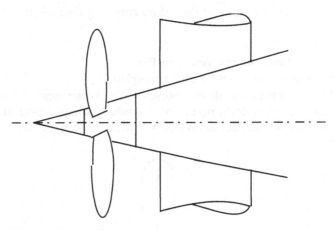

Fig. 6.10 Single axial propeller

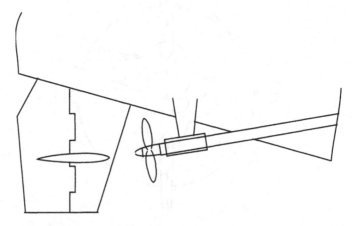

Fig. 6.11 Ship-like twin screw propulsion

However, the single axial screw propeller (Figure 6.10) requires the transmission of all the propulsion power through a single shaft, which may pose machinery problems. It also means that any damage to the propeller or shafting deprives the submarine of its propulsive capability. Therefore the designer might, in the context of the overall design, choose to pay a propulsive efficiency penalty and adopt a twin or even a multi-shaft arrangement. (Figure 6.11) In that case, changes to the shape of the stern may help improve the wake conditions, while the large wake variation due to the bridge fin shadow can be avoided.

Alternative propulsors

6.18   However, the single screw propeller is not the only or necessarily the best way in which good propulsive efficiency can be achieved with an axisymmetric geometry. Although efficiency can be largely accounted for in terms of axial acceleration of the fluid, the action of a screw propeller also imparts rotational motion or 'swirl' to the fluid downstream, which constitutes a waste of energy. The design of a large diameter propeller also leads to the requirement for high torque on the shaft and low speed of rotation (rpm), which not only poses problems for the propulsion machinery design but also requires that the hull resist the torque reaction of the propeller.

One way of recovering the lost energy due to swirl and balancing the torque, is by use of co-axial contra-rotating propellers (Figure 6.12 (a)). The combined action of the double propellers cancels out the rotational energy loss and so a higher propulsor efficiency can be obtained. At the same time the opposite rotations of the two sets of blades cancel out the

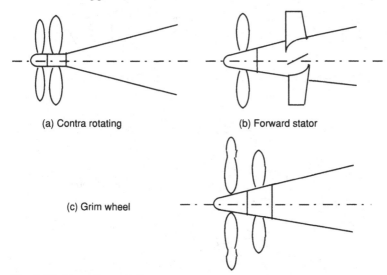

(a) Contra rotating   (b) Forward stator

(c) Grim wheel

Fig. 6.12 Propeller variations

torque reaction. However, there is a substantial penalty of complexity in the design of the co-axial shafting with its bearings, the shaft/hull pressure seal and the arrangements inboard to drive the two shafts.

A somewhat simpler solution is the use of stator blading in conjunction with the rotating propeller (Figure 6.12 (b)). Stator blading may be mounted on the hull ahead of the propeller and angled to produce pre-swirl into the propeller so that there is little or no swirl downstream of the propeller. More difficult to fit are stator blades behind the propeller to straighten the flow and produce additional thrust and counter torque, the difficulty being to provide support to the stator blades. One device which has been tried on surface ships is known as the 'Grim Wheel', which has blades rotating freely on the hub of the propeller; over the propeller disc the blades effectively act as a turbine, taking out the rotational flow, while the tips act as additional thruster blades (Figure 6.12 (c)).

Another alternative to the open screw propeller is the ducted propulsor, which has a cylindrical shroud or duct around the rotor. (Figure 6.13) The duct affords some protection to the rotor blading against surface debris or under ice. The action of the duct is, however, much more than this. In an open propeller, the tips of the blades cannot provide much thrust or lift because the pressure difference between the front and

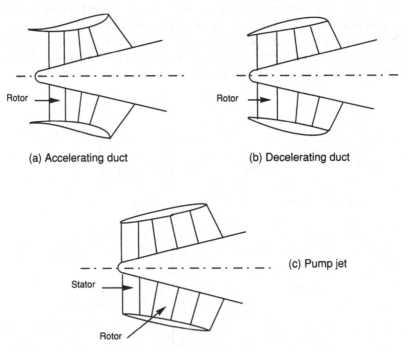

(a) Accelerating duct   (b) Decelerating duct

(c) Pump jet

Fig. 6.13 Duct variations

back of a blade causes the flow to spill over the end, resulting in a rotation of the fluid leading to what is called a 'tip vortex' downstream. Quite apart from the thrust deficiency at the blade tip, the vortex can lead to cavitation in near surface operations due to very low pressure at the core of the vortex, which is a highly undesirable noise making process.

If the blading is designed to run with a small clearance from the duct the pressure difference between the two faces of the blade can be sustained. This is one reason why a smaller diameter ducted propulsor can be used compared with an open screw propeller; and it is also able to make fuller use of the higher wake effect due to the smaller diameter. The duct can also be designed to control the flow at the rotor. If it is designed with a cross-sectional area greater at the rotor than at the inlet or outlet it reduces the flow velocity and increases the pressure at the rotor blading, whichs inhibits the onset of cavitation in shallow water. However, there is a penalty because a duct of that shape develops drag in addition to its form drag, so the rotor has to produce more thrust (Figure 6.13 (b)).

Alternatively the duct may be designed to accelerate the flow into the rotor (Figure 6.13(a)). It can be shown that such a duct develops some of the total thrust itself, reducing the loading on the rotor. For high thrust conditions this can give greater efficiency than an open propeller. The use of a ducted shroud enables the stator blading to be located ahead of or abaft the rotor or both. The stator blading can take out the swirl as already discussed, and can be used to modify the flow through the rotor, including smoothing the variations in wake velocity. Such a device is known as a Pump Jet (Figure 6.13 (c)). It presents a complex design problem in detail, but provides hydrodynamic and acoustic advantages where high power is to be transmitted. It has some disadvantages hydrodynamically in providing less dynamic stability and much poorer astern performance. It is also a very heavy component at the very stern of the vessel, and due to its complex construction it is considerably more costly than an equivalent propeller for the same duty.

Other propulsive devices

6.19 Whilst the foregoing discussion covers the main choice of propulsor for submarines some other alternatives are worth mentioning.

In small submersibles the requirement is likely to be not so much for high speed as for very close attitude control, slow or stopped. Thrusters for that purpose can be small propellers (ducted or open) mounted with their drive motors in rotatable pods, so that vectored thrust can be obtained. The thrusters can also serve for propulsion by directing the thrust rearwards. To achieve full six degrees of freedom control a number of pods are required (Figure 6.14).

For vessels with a requirement for position/attitude control as well as for speed, the control thrusters may be enclosed in lateral ducts across the hull similar to surface ship bow thrusters. Such an arrangement is used for the Deep Submergence Rescue Vessel, which requires attitude control for mating with the hatches of a crippled submarine. It is possible to enclose the main propulsion thrusters in longitudinal ducts, but generally much lower efficiencies result due to duct piping losses. However, it is conceivable that some form of boundary layer control could be achieved by appropriate location of the inlets.

A more unconventional propulsor was suggested by Hazelton, which consists of two large rings of blading sited fore and aft outside the hull. The blading is controlled cyclically similar to helicopter rotors, so that forward thrust can be developed with some boundary layer control, while lateral thrust and couples can be generated to change the attitude of the submarine.

There are many other unusual methods of propulsion, some mimicking fish fins, but the one of particular interest is magneto-hydrodynamics (MHD). This is based on Maxwell's theory that there is an orthogonal relationship between magnetic field, electric current and motion in a conducting fluid. Sea water is a reasonable conducting fluid so that if a magnetic field can be set up and an electric current field generated at

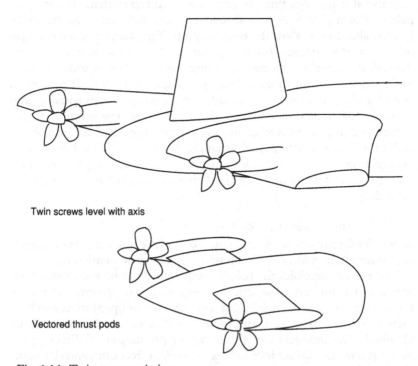

Twin screws level with axis

Vectored thrust pods

Fig. 6.14 Twin screw variations

right angles there will be flow of water perpendicular to both fields. The device may be designed within a separate duct or pod mounted on the hull or the fields may be generated just outside the hull itself, both inducing astern flow and thus thrust for propulsion. By careful design of the fields complete control of the flow over the hull might be achieved, reducing the viscous resistance. However, the MHD device calls for very large magnetic fields and currents, requiring large generating power. For the military submarine there is also the danger of creating a large magnetic signature that could make detection all too easy.

The propelled vehicle

6.20   The traditional naval architecture approach to resistance and propulsion in design has been described in the preceding sections. The procedure aims to obtain a low resistance hull form (effectively towed resistance), to identify a suitable optimum propulsor to provide the required thrust, and to take account of interaction effects between the hull and propulsor. Although this is a well-established procedure, it does raise the question as to whether the design solution obtained really is the optimum result. It is conceivable that as a total propelled body there are other optimal solutions.

The reason for the slowly tapering after body of the 'ideal' form is that this shape delays or averts the rapid increase in boundary layer thickness and possible separation of flow as the velocity over the body decreases and the pressure increases towards the after end. The action of the propulsor at the stern is to accelerate the flow towards it, and this influence extends one or two diameters ahead of the propulsor. It is arguable that if the propulsor and body could be merged so that the acceleration effect of the propulsor cancels the deceleration effects of the body (a form of boundary layer control) then another solution to the total propulsion problem might be available. Thus a body shape could be postulated in which a slowly increasing or parallel cylindrical part extended almost to the stern, where there would be a rapid decrease to the after end. The propulsor would then be sited close to the stern so that its acceleration effect cancelled the separation which would otherwise occur. That configuration could not only have hydrodynamic advantages but also provide good internal space and hydrostatic advantages by having more volume in the after body. There is some evidence to support the concept in what is known as the Griffiths aerofoil and similar configurations tested in torpedoes. The problem is that, whereas the torpedo can be tested full scale, it would be a bold step even if model tests supported the idea to build a full scale submarine with such a novel configuration and many engineering problems with the propulsor and control systems would have to be solved. As we have previously commented, considerable research and testing support would be necessary before such a radically novel design could be accepted.

DESIGN ASPECTS OF PROPULSION PLANTS

Sizing the propulsion plant

6.21 Having described the steps taken in deciding on the configuration of hull and propulsor to minimise the power requirements, we now address the considerations which bear on sizing the power plant. Because the propulsion plant occupies around 50% of the pressure hull volume, it has indeed an important role in sizing the vessel. (Figure 6.15)

For the purposes of discussion it will be assumed that the power plant is of the standard diesel electric type. Later alternative forms of plant including nuclear will be briefly discussed.

There are essentially four components of a diesel electric power plant. Each involves somewhat different criteria in sizing, so although the issue has necessarily to be viewed as a whole, with the components interdependent, there is scope for variation in the sizing of the components which merits regarding them individually.

These components are:
      (a)   Propulsion motors,
      (b)   Batteries,
      (c)   Diesel generators,
      (d)   Fuel,
and we go on to discuss them in turn.

Propulsion motors

6.22 These are direct current electric motors, usually arranged with the rotor directly coupled to the propeller shaft. They are required to be of a size to deliver the necessary shaft power at the top speed of the submarine. However, the selection of a suitable motor is not only governed by the maximum shaft power or the voltage available to drive it. Because it is directly on the propeller shaft its speed of rotation (rpm) is that at which the propeller has been designed to deliver full thrust. Similarly the torque output of the motor must match the torque of the propeller at full power conditions. Hence there is a matching problem to be considered.

We have shown that high propulsive efficiency calls for a large diame-

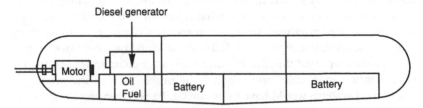

Fig. 6.15 Propulsion elements inside hull

ter, low rpm propeller with high torque. It is not our intention here to go into details of electrical design, which are touched upon in a later chapter. Suffice it to say for now that there are limitations on the field in the gap between rotor and stator of an electric motor and the circumferential force that can be generated. Thus a compact design of electric motor for a given power would give only low torque at high rpm. The requirement of the propulsor for high torque at low rpm is directly in opposition to this. The propulsion needs call for a large diameter rotor, resulting in a heavy, volume-consuming propulsion motor at the narrow part of the pressure hull stern, and so posing hydrostatic balance and space layout problems which are difficult to satisfy. There may in consequence have to be a compromise in the overall design in which the propulsor rpm are increased and the torque decreased to match a motor of tolerable size. As we explain in the final chapter there may not be a continuous range of choice, but rather step jumps in the size of available motors, which impact on the propulsor design. A possible alternative is the use of two smaller motors on the shaft, which trades diameter demand for additional length demand, though it also complicates the electric control. That configuration might be adopted because it provides some redundancy to the propulsive system and because it permits changes in propulsion at lower speeds, for although sized for full power a propulsion motor will spend most of its working life at much lower powers and rpm.

The designer could also choose indirect drive of a motor through a gearbox to the shaft, which overcomes the matching difficulties but at the expense of cost, weight and space of the gearbox. It is not favoured for submarines because a direct drive system can be made very quiet, whereas it is difficult to eliminate noise from a gear box.

Associated with the motor itself will be a fairly large switchboard that has to deal with high currents and voltages for various power settings.

Recent developments in the design of permanent magnet motors may provide a solution to some of these problems described above. Motors of this type offer the prospect of smaller size capable of matching the propulsor requirements with a considerable reduction in switchgear to provide control, which would not only reduce space demands but also offer an improvement in reliability and maintenance.

Batteries

6.23   In the fully submerged condition it is the stored electrical energy of the batteries that meets the power demands of the vessel. As well as providing the propulsive power, the batteries also have to provide power for the operation of sensors, weapons and auxiliary machinery and those for habitability, ventilation and air-conditioning of the crew. They constitute a steady drain on the batteries, governed by the length of time submerged, and together are termed the Hotel Load. For estimating purposes the Hotel Load is taken as a constant, even though it is reduced for ultra-quiet or sustained submergence by switching off all non-essential equipment, and even limiting the running of essential equipment.

The main demand, which calls for a large number of battery cells, is that for propulsion. Even a large number of cells will be drained very rapidly at top speed, lasting perhaps only 30 minutes. For that reason, it is not usual to size the batteries on a full power criterion, but to base the capacity requirement for the battery either on the criterion of a single cruise speed submerged multiplied by submerged time, or on one related to an operational pattern of speed submerged. Having determined the capacity for whichever of those criteria is selected, the length of time at top speed with initially fully charged batteries can be deduced and it is then adopted as a design specification for the boat. As the propulsive power is proportional to the cube of speed, extended submerged time necessitates acceptance of fairly low speeds. If, however, it is the distance travelled submerged which is regarded as operationally important, there will be an optimum speed of transit below which the Hotel Load will drain the batteries before the required distance can be achieved.

As before, there are in battery capacity determination step jumps rather than a continuum of choice. The individual cells are linked in sections to provide an operational voltage, e.g. 220V DC from approximately 100 cells. Two such sections can provide either 220V or 440V and four sections 220V, 440V, 880V. Hence to some extent the choice of standard battery size is fairly closely constrained.

Diesel Generators

6.24 The diesels can only be run while connected to the atmosphere, i.e. either surfaced or snorting. Their primary purpose is to re-charge the batteries which are drained during submerged running. When doing this they also have to supply power to the Hotel Load and power to propel the vessel at snort or surface speed. The battery charging power is governed by the capacity of the batteries (already determined) and the maximum charging current and top-up current which the battery cells can accept. Thus the time to charge the batteries, and hence snorting time, is governed by the batteries not the diesels. This assumes the diesels are at least capable of meeting the maximum charge rate. The Hotel Load is essentially fixed, so the sizing of diesel generator capacity over and above this is a question of snort (or surface) speed and power. For reasons of depth control and the forces on masts and periscopes, snort speed is usually limited to 10 kts. Therefore the transit distance in the snort condition is governed by charging time and that order of speed.

A factor considered of operational importance is what is known as 'the indiscretion ratio', i.e. the ratio of time snorting to time submerged. It can be seen that this is mainly governed by the submerged speed pattern and is not directly influenced by diesel power selection.

Having assessed the requirement for diesel generator power there is the selection of the engines, which again tend to come in discrete steps. It is not usual to accept only one diesel engine, as this would mean that any breakdown or fault would render the submarine totally incapable of recharging its batteries. Two diesels are more common but there are arguments related to redundancy and maintenance which favour three or four engines. Basically the argument is that if it is assumed that one engine fails, the vessel should still be able to recharge batteries though at a slower speed.

Fuel

6.25   The sizing of the fuel stowage relates to all the foregoing. If the possibility of a submarine leaving harbour fully charged and returning with fully discharged battery is ignored, then essentially the fuel stowage has to meet the total energy requirements of the patrol. The charging of batteries is essentially an intermediate transfer of the energy of fuel by combustion to a stowage system for use when air/fuel combustion cannot be accomplished. Thus the fuel capacity is sized by the Hotel Load for the total patrol period plus the total propulsive energy of the patrol. The latter may be simply assessed by assuming a total distance/range of the patrol and an average speed of advance. A more detailed calculation may be made by assessing the time and speed surfaced, time at snorting speed and the time and speed pattern whilst submerged.

Having assessed the total consumed energy requirement, the fuel requirement is deduced from the various energy conversion efficiencies of the systems and the specific energy content of the fuel.

Nuclear power

6.26   It can be seen that the conventional submarine is severely limited in its submerged capability by the energy storage system, i.e. the battery capacity which is carried on board. That limitation was almost entirely overcome when a practical nuclear reactor plant was developed, as it effectively gives an almost infinite amount of energy storage, reckoned not in hours, not even by a single patrol, but many years of power availability (though the rate at which the power can be drawn from a reactor is limited by design factors appropriate to the particular reactor).

One of the problems with a reactor plant of the Pressurised Water type (the PWR) commonly used in nuclear submarines is that there is a minimum size at which such a system can be built and operated. Not only does it require a massive primary circuit, but with the shielding and protection necessary with such a system the totality is a very large unit of equipment which cannot be readily varied either in power output or in volume and weight demands on the vessel. At the present state of technology this virtually excludes the consideration of a PWR plant for anything less than a 3 500 to 4 000 tonne submarine. A nuclear submarine is both expensive in first cost and to maintain, run and crew. It is not considered a feasible proposition by many Navies throughout the world.

The energy from the nuclear reactor of a PWR plant is converted into power by means of an indirect steam system, providing direct propulsive power to steam turbines through a gear box to the shaft and electrical power via turbo-generators. It is still usual to fit an electric propulsion motor on the shaft but as a fall-back, low power drive, not the primary propulsion system of the vessel as is the case for the conventional submarine.

High capacity batteries

6.27   Because of the considerable disadvantages of a nuclear power plant except for large navies able to operate large submarines, technological development is returning to tackle the problem of improved energy storage and power availability by other means for a vessel fully submerged beneath the surface. Since the batteries in a conventional vessel take up a considerable amount of the weight and space one quest is for higher capacity batteries storing energy at a higher rate of watts per litre or kilogram than current lead/acid batteries are capable of doing. There is some progress here which suggests that perhaps three times the power in terms of energy storage could be achieved through other forms of chemical batteries. Such a development would considerably enhance the submerged performance characteristics of the conventional submarine. One such battery is the Lithium Aluminium/Iron Sulphite high temperature battery, under development in the UK, though there is some penalty in providing thermal insulation around the battery.

Air independent power systems

6.28   Another method of providing power in the submerged condition is by what are known as air independent power (AIP) systems. A conventional diesel engine power plant converts the energy in a diesel fuel oil by its oxidation with oxygen from the air drawn in; the air provides not only the oxygen for combustion but also a fluid medium on which the engine runs. When submerged the oxygen and fluid medium are no longer available to the submarine, so a conventional power plant is impractical for use except for extremely limited periods. Those circumstances can occur when snorting, when the induction head dips below the surface and causes an automatic shut-off in order to avoid the ingress of water down the intake pipe, so that momentarily the diesel engines are running on the air contained within the submarine. In a very short time a vacuum is created in the boat which would be dangerous to the crew and not very good for the diesels either, and so the diesels have to be shut down if the snort induction mast does not quickly re-emerge above the water surface. One solution to the submerged problem would be to carry not only the combustible fuel but also the oxygenation agent on the submarine, i.e. carry oxygen in some form which may either be as liquid oxygen or as a highly oxygenated fluid of which high test peroxide is a well known form. As mentioned in Chapter 2, in the latter part of World War II the Germans developed a submarine running on high test peroxide using the Walther turbine propulsion plant which involved the external combustion of fuel and oxygen to provide the heat source for the boilers. The plant gave the capacity for quite high speeds, although the endurance was still limited to a few hours by the need to carry both the high test peroxide and fuel on board the vessel. Further tests on similar vessels in the Royal Navy after the war were discontinued with the advent of the nuclear plant. One other factor leading to that outcome was the extreme danger in the use of high test peroxide which will readily burn with virtually any other material with which it comes in contact. Nevertheless, if the high security, safety and cleanliness standards which have to be adopted in the nuclear plant were applied to peroxide plant then the same degree of safety would be conferred on such systems.

A similar method is adopted in the use of the Stirling Cycle with a reciprocating engine, which again uses an external combustion system to provide a heat source running on a heat engine cycle. A plant of that type has been fully developed in Sweden and is now being adopted on some submarines, though its capacity for high power has not yet been developed and it is primarily a means of providing a sustained endurance at moderate speeds. Another alternative is the use of what is known as the Re-cycle Diesel Engine; in that application a normal diesel engine is operated on a closed circuit by circulating the same primary gas fluid around the system, but with additional oxygen injected into the fluid every cycle to enable combustion of the fuel within the cylinders. With such a system it is necessary to remove the resultant carbon dioxide/monoxide from the exhaust system.

An AIP development of much promise is the fuel cell, which has particular advantage for submarine application because the technology it uses provides an output of DC electricity from inputs of oxygen and hydrogen in one form or another by an electrochemical process that involves no direct mechanical action; it is inherently silent, pollution free and of good efficiency (around 50%). Six or so different kinds of fuel cell have been devised up to now, differentiated in the main by the substance forming the electrolyte and the operating temperature (typically ranging from 50°/70°C to 900°/1000°C). Some fuel cells lend themselves to use in vehicles at power levels currently in the order of tens of kilowatts, while others are intended for use for land-based power generation purposes at power levels currently in the order of hundreds of kilowatts. For submarine applications interest clearly lies in mobile fuel cell systems, but as things stand at present they represent a potential energy source rather than one actually available for installation as a proven and reliable plant. The policy being followed by the navies of several countries is to keep abreast of commercial research and development into mobile fuel cell plants, with a view to taking up the option to adapt a suitable plant for submarine use at an appropriate stage.

If any of the systems were adopted or developed in the future the main drawback is likely to be the need for additional volume to be provided aboard the submarine in order to store the fuel and oxygen or alternative chemical agents.

DESIGN ASPECTS OF POWERING

6.29   Looking at the powering of submarines from the overall design aspect, it is a significant driver of size not only because the propulsion plant occupies so much internal space, but also because the power/size relationship is of a self-perpetuating nature, i.e. the more power that is sought for, the larger the submarine has to be, which necessitates yet further power, and so on. It consequently behoves the designer to press arguments for containing operational ambitions for high underwater speed (in view of the cube law characteristic of the power/speed relationship) as well as aiming to achieve a form and propulsor arrangement suited to good propulsive efficiency.

It is with regard to the selection of form that powering is an important influence on the designer. For, as we have seen, although other considerations may well cause him to depart quite a way from the 'ideal' form, it is concern about the powering consequences which sets boundaries to the extent of departure.

7 GEOMETRIC FORM AND ARRANGEMENTS

INTRODUCTION

7.1 In preceding chapters we have touched in a rather piecemeal way on a number of considerations which bear on the choice of geometric form of a submarine and on the way in which its contents, both inside and outside the pressure hull, could be arranged to best effect. Our purpose in this chapter is to collect those considerations together and also to introduce several other issues associated with arranging the contents of the submarine.

In dealing with the matter of arrangements we will be addressing an aspect of the activities of the submarine designer which are what one might call architectural by nature. As we are both naval architects by profession we can take that description for granted our regret is that so much of a naval architect's time is taken up with engineering matters that he can be limited in the attention he is able to give to architecture. Yet it is an area of activity in which there is still scope as well as need for some art amongst so much science.

It might appear at first sight that the form and arrangements of a submarine present much simpler problems than those of a surface ship. Essentially a submarine is a long tube in which the disposition of most of its contents is arranged longitudinally with little scope for vertical variations. In a surface ship, on the other hand, with its multi-deck configuration, there is freedom for spatial interaction longitudinally, vertically and athwartships to be taken into account. But because the constraints in submarine design are tight indeed, the entire available volume being confined to the submerged form, its efficient utilisation for layout purposes is in fact a demanding exercise calling for skill and ingenuity.

FACTORS INFLUENCING FORM AND ARRANGEMENT

The shape of the outer envelope of the submarine is, as we have seen, selected to give a form of minimum resistance to forward motion consistent with other demands and as we show in the next chapter selection of the form is also influenced by, and influences, the control dynamics of the boat.

Within the outer envelope the pressure hull is contrived to provide an efficient structure to withstand pressure and also to contain the majority of equipment and spaces required in the submarine. The requirement for an efficient structural configuration for the pressure hull is not always entirely compatible with a streamlined outer envelope for low resistance; one ideally requires constantly changing shape along its length whilst the other requires constant diameter. The differences between the outer envelope and the inner pressure hull forms provide space for some external stowage of equipment but primarily for tankage external to the hull. What is left when those requirements have been satisfied has to be free flood space which is an undesirable feature in the overall design as it is useless volume which still has to be propelled. The space between the inner and outer boundaries has to be of such dimensions as to be workable in terms of structural building and accessibility for painting and surveying and maintenance at later stages. It is quite possible that a pressure hull fully enclosed within an outer form may lead to far more external volume than is needed for external tankage in the vessel. Thus the simple wrapround configuration is not usually a very efficient design for a submarine, which leads designers to adopt different configurations as regards the utilisation of the space between the inner and outer hulls. In many modern designs of the tear-drop form, the pressure hull forms a high proportion of the outer surface envelope of the vessel, and free flood and tankage space is provided only at the forward and after ends. However, such an arrangement can lead to a shortfall in the required external volume requirements of the vessel, and also to a longitudinal imbalance between the submerged and surface conditions.

Faced with that dilemma the designer has several options (Figure 3.4):

(a) He could lengthen the submarine either forward or aft in order to increase the total volume and correct the longitudinal balance, though in so doing he would depart from the ideal length to diameter ratio;

(b) He could add tanks along the length of the pressure hull in the form of blister or saddle tanks to increase the external tankage space, though this would cause the external form to depart from the ideal axisymmetric shape and increase drag;

or

(c) He could include internal tankage for volumes which might otherwise have been located external to the hull; that option has been suggested earlier as a means of adding to the reserve of buoyancy by having some of the MBTs inside the pressure hull.

There is, however, an even more radical approach which a designer could resort to. He may adopt a more complex geometry for the pressure hull in order to generate extra volume between the inner and outer envelopes. This could be done by reducing the diameter of the pressure hull over selected parts of the hull. Typically it may be reduced over the forward length of the hull to increase the volume of external tanks forward, and reduced aft to increase tankage there. The diameter may also be reduced over the mid-length area causing a waisted section of the pressure hull. This causes complications both in the construction of the pressure hull and the stresses it experiences on diving. The approach is feasible only if, along the length of pressure hull, there are certain internal spaces which do not require the full diameter of the pressure hull. Such spaces do occur forward where the torpedo compartment is located, amidships in the region of the auxiliary machinery spaces, or aft where the space needs are mainly for the shafting with its equipment and propulsion motor.

FACTORS GOVERNING DIAMETER OF HULL

7.3   The number of decks that can be accommodated within a pressure hull is a function of its diameter. The recognition that the distance between decks needs to be slightly larger than the height of the average man plus passing services has led to the proposition that pressure hull diameters come in unique steps. (Figure 7.1)

In very small submarines this means that the diameter of the hull is approximately the height of a man and the deck has to be the lower surface of the hull. To introduce a single through deck the diameter needs to be slightly more than twice the height of a man, leading to the next step of something between a five and six metre diameter hull, with the main deck at mid depth and compartments above the deck governed by the overhead curvature of the hull and compartments below the deck governed by the underneath curvature of the hull. The next step in diameter involves adding a further deck height, so that the pressure hull diameter is now in the order of seven and half metres. Yet another deck leads to a diameter common to many of the large submarines presently in ser-

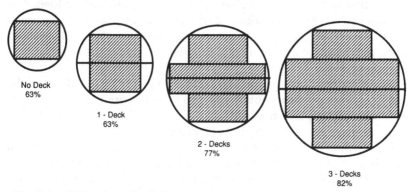

No Deck
63%

1 - Deck
63%

2 - Decks
77%

3 - Decks
82%

Fig. 7.1   Utilisation of hull diameter

vice of about ten metres. The argument for discrete steps is that to adopt intermediate diameters would lead to wasted volume or unusable spaces in the submarine, either by giving too much overhead space in upper compartments or too little lower space in bottom compartments. If discrete steps in pressure hull diameter are adopted this can limit the ability to achieve an appropriate length to diameter ratio for powering purposes, but while that aspect gives useful guidance in the choice of dimensions of the main pressure hull it cannot be regarded as overriding. In any case, although the arrangements with pressure hulls of intermediate diameter can be awkward, they are not necessarily entirely unworkable. It is quite possible that the odd spaces occurring with intermediate diameters can be used either for passing services, which would otherwise interfere with other equipment and cause changes in longitudinal disposition, or can be accommodated in the lower part of the hull with higher tank spaces or better space for auxiliary machinery and battery tanks. It should also be appreciated that the foregoing considerations really only apply to the main block of accommodation and operational spaces, where equipment and men dictate the deck height. Standard deck height dimensions do not apply to machinery spaces nor to the weapon spaces at the fore end of the vessel. The spaces usually have their own headroom dimensions; for instance, it is usual to provide sufficient clearance over the diesel engines to allow overhauls, including lifting the cylinder heads, *in situ*.

INTERNAL ARRANGEMENTS

7.4   The arrangement of compartments within the pressure hull entails not only the satisfaction of the demands for the amounts of volume involved but also some geometric considerations of the shape of those volumes and their longitudinal or vertical disposition. Thus, for instance, the propulsion plant generally needs to be aft and aligned to link up with the line of shafting, thrust block, gearing and motors. Similarly at the fore end the torpedo compartment needs to be adjacent to the discharge system and aligned so that weapons can be loaded in and out of the tubes. The trim and compensation tanks can be arranged in the lower part of the hull but also require appropriate disposition along the length of the submarine in order to achieve the required corrections of trim.

Heavy weights, for example the batteries, need to be low down in the hull in order to provide stability and also to be disposed to maintain longitudinal balance, which effectively determines the location of battery compartments in a conventional submarine.

Sub-division of compartments by transverse bulkheads is not always a free choice based on the internal arrangements since, as we have seen, the position of bulkheads is also influenced by structural considerations of compartment length in relation to overall collapse of the pressure hull.

Some of the compartments within the vessel have requirements for juxtaposition. Thus the control room, communications office, sonar office and navigational and control area need to be associated together. It is normal for the periscopes and masts to be in this area, thereby linking the siting of those internal compartments to the bridge fin. Similarly, the accommodation of the crew, as regards sleeping areas, galleys and dining rooms and bathrooms, needs to be arranged in some suitable fashion to provide for convenient living on board. Associated with the living spaces is the need for waste disposal tanks fairly close underneath the bathrooms and galley, and also food stores positioned where they are readily accessible from the galley. Another requirement is for escape compartments to be situated at either end of the vessel and associated with access hatches. Apart from escape, hatches are usually associated with other functions, there being a need for hatches for storing the vessel and machinery removal plus access to the bridge conning position. As this could lead to a large number of hatches if dealt with individually, attention has to be given to combining functions so that as few large penetrations as possible are made in the pressure hull.

A useful aid to checking the disposition of all internal pressure hull volume to best effect is the so-called Flounder Diagram. (Figure 7.2) The outer boundary of this diagram is a curve of cross-sectional areas of the pressure hull to a base of length, which represents the longitudinal disposition of available volume. Space demands can be indicated as individual areas in the diagram, together with their longitudinal and vertical location demonstrating how the available internal space has been used up. External space demands can also be shown in the same diagram.

Although the diagram gives a two-dimensional representation of volume distribution it does not illustrate the shape of the volume and hence its utility, and so some care has to be exercised in filling in the diagram. A modification to the diagram is to show a second curve of the volume available in terms of deck area with full headroom which indicates the disposable volume for major compartments. The spare volume can be assigned to system runs, storage and tankage which do not necessarily require clear headroom for access.

As well as the individual demands of compartments and operational spaces and their inter-relationship, it is necessary to consider the general movement of crew through the boat. It is important that this movement should not interrupt activities in the operational spaces. In older vessels, however, interruption was unavoidable as the control room was situated in the middle of the vessel so that crew going on watch to aft machinery spaces from forward accommodation had to pass through the control room often just at the time of action stations. Even more unfortunate circumstances could arise such as the siting of the galley aft of the control room when the mess spaces were forward, so that the control room

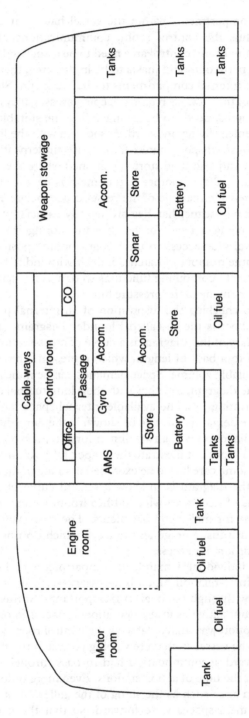

Fig. 7.2 Typical Flounder diagram

became a passageway for cook and mess men carrying trays of food and soup through that vital compartment. Especially for submarines which only have a single through deck it is worth considering an arrangement in which the propulsion and machinery is aft, the accommodation is sited amidships and the command, control and weapons spaces are forward (Figure 7.3) which allows a rapid change of crew on watch without interfering with the operation of the vessel. Much more freedom for arrangements is available with the multi-deck arrangements in the larger diameter submarines.

SOME DETAILED CONSIDERATIONS

Siting of acoustic systems or sonar

7.5   The primary sensors for an underwater craft are those of its acoustic systems or sonars, and so they merit priority in the provision of prime sites to enable them to function at their best. For both military and commercial vehicles the major component of sensor equipment will be

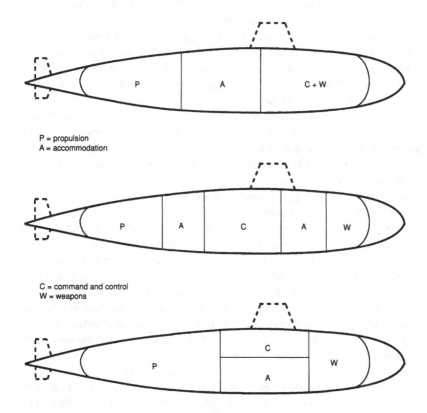

P = propulsion
A = accommodation

C = command and control
W = weapons

Fig. 7.3   Different arrangements of major space blocks

the bow sonar, which is usually required not only to have a good forward looking sector of view but also as far round to the sides as possible, with some arcs towards the stern. The dependency of the vehicle on the effectiveness of its acoustic devices means that the bow configuration must be contrived to suit the sonar equipment as far as possible, even if this requires the form of the hull at the bow to depart from the ideal form for resistance and hydrodynamics. Thus on many submarines the bow shape is different from the ideal elliptical form of bow, which results in an unsymmetrical form above and below the axis. However, despite this effort to afford the sonar very good acoustic conditions, one factor which may degrade sonar performance is the water flow over the area in which the sonar acts. A highly unsymmetrical bow in which there can be complex and possibly turbulent flows may degrade the sonar considerably. Thus some sonars are embedded within the outer envelope of the hull and and covered over with an acoustic window of an ideal hydrodynamic form in order to provide more favourable conditions for the sonar. As we shall see later, there are other contending demands for space in the bow of the submarine, but in nearly all designs these are treated as secondary to the interests of the sonar equipment.

The more recent requirement for longer range and lower frequency sonars has led to the development of flank arrays which demand both length and depth in such a way that the whole of the submarine hull has to serve as a platform for the flank array. This demand may necessitate a greater length of the hull being constant diameter despite the hydrodynamic requirement for changing sections. Also, if the pressure hull forms the outer boundary of the hull then the flank sonar array has to be mounted outside the hull. This results in a bulge on each side of the hull where the array is fitted, which will add to the drag of the vehicle. However, it is not only the increased drag of the vehicle which is important but also the flow disturbance caused by the bulges on an otherwise smooth hull. Turbulent noise thereby generated may cause interference and degrade the effectiveness of the sonar. It is therefore imperative that, as far as possible, flank arrays should be incorporated into a smooth hull so that the flow turbulence is kept to a minimum. This need provides a strong argument for adopting the wrapround form in order to give space into which the sonar can be recessed. A consideration that is sometimes overlooked is that flank arrays have to be accurately aligned in order to analyse the incoming signal, but are mounted on a hull which is flexible. It is not possible for precise alignment to be retained in the various loading conditions of the submarine and so the software has to incorporate means of detecting and correcting for the flexure that occurs.

Other sonar intercept sets use relatively small transmitter/receiver transducers but require them to be spaced out over a long baseline. The need to provide this baseline may lead to the design of a submarine longer than would be determined from hydrodynamic considerations. If increased length cannot be provided, some degradation of the performance of the sonar would have to be accepted. In these sonars also it is important to devise software which can check the alignment of the individual transmitters/receivers when the vessel flexes under changes in load. Siting of other intercept sonars and underwater communications systems can usually be treated as a matter of convenience. Thus a sternward-looking sonar might conveniently be sited in the rear part of the bridge fin as this gives a reasonable look astern view over the main pressure hull. For the same reason, sonar sets might be fitted in the forward or after parts of a ballast keel if this were fitted.

There is an increasing use nowadays of towed systems both for sonar and communications. These need to be provided either with hook-on points or, preferably, stowage where they can be reeled in and stowed when not in use. The latter calls for space either in the stern of the vessel or in the casing aft. Another important feature in the arrangement of such systems is that they should be trailed from a point which avoids the possibility that the cable may be inadvertently drawn into the propeller, either in a turn or due to acceleration/deceleration of the vessel, thereby damaging or cutting off the cable.

Because sonar equipment depends upon a good acoustic environment in which to operate, it is important to try to avoid the possibility of inboard noises and noise makers such as machinery and pumps transmitting these interfering sounds into the sonars, obscuring their ability to hear incoming sounds. Two modes of transmission of onboard noises can occur: one is through direct structural transmission from the noise sources to the sonar sets, while in the other the sound is transmitted into the water from the onboard sources and thereby reaches the sonar via water-borne paths. Both modes can lead to false echoes and degradation or blanking of certain arcs of view of the sonar. Every effort needs to be made to site sonar equipment as remotely as possible from potential noise makers, and then to do the best possible to reduce self noise by damping and isolating transmission paths between the noise makers and the sonar sets. Where very powerful active sonar sets are employed, care must be exercised to ensure that an excessive noise level does not penetrate into the boat. The high sound pressure level associated with such sets can be damaging to the crew and therefore steps may be necessary to acoustically isolate compartments in which the crew live and operate.

At the concept design stage it is not easy to assess the effects of such sound attenuation features on the initial sizing of the hull and yet provision has to be made in budgetary allocations for an accurate assessment to be made at the detailed design stage. This emphasises the need for good data bases as an aid to estimation. Where possible, weights and space density factors should be used from previous designs of submarine in which a corresponding standard of sound isolation was provided. If that data is not available, the designer must make a specific allowance for implementation of sound attenuating measures and consider increasing design margins accordingly. Allowance for sound isolation measures is not only a matter of catering for their weight as there may be substantial increases in space demands for the incorporation of more elaborate monitoring systems.

Siting of torpedo tubes

7.6   The traditional siting of torpedo tubes was at the bow of the submarine, which meant that to fire a torpedo at a target the submarine would be pointed towards the target. With modern guidance systems in underwater weapons pointing is no longer necessary, but nevertheless the most favoured position for torpedo tubes is still that at the fore end of the submarine. That location does, however, cause some clash of priorities between the interests of the torpedo tubes and those of the bow sonar which is likely to be the fire control sonar for the weapons system.

There is a difficulty of physical geometric interference between the two systems and also an acoustic difficulty. The discharge of a torpedo is a noisy process from the moment at which the bow shutters and bow caps are opened, followed by the impulsive firing of the torpedo out of the tube. It is clearly undesirable for this process to be taking place very close to the main sonar set involved in fire control. However, there is little room at the bow of the submarine to achieve much separation of the two systems. An arrangement commonly adopted is to mount the sonar high up in what might be called the forehead position at the bow, faired into the fore end of the casing or superstructure, which leaves the lower or chin position of the bow free for torpedo tubes. (Figure 7.4) The torpedo tubes

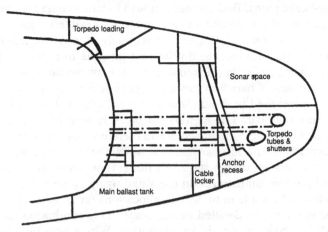

Fig. 7.4   Typical fore-ends arrangement

will then penetrate the foreward dome bulkhead in the lower part of the pressure hull circle, which can cause problems because the space inboard of the tubes has to be kept clear in order to align weapons for loading into the tubes. There may also be little room underneath for the associated tankage and forward trim tanks required. On the other hand, having the torpedo tubes in the lower part of the bow allows for wire guidance torpedos to trail their guidance wire from the lower chin position at the bow and trail under the hull without interference with other parts of the boat. This arrangement also raises questions as to what to put in the upper part of the pressure hull above the torpedo stowage position in line with the tubes. With a large number of weapons, the whole of the pressure hull may be given over to weapons stowage, in which case there would be little difficulty except that weapons have to be lowered down to align them with the tubes when selecting from the choice of weapons. If, however, only lower stowage is required then there would be space above the weapons stowage compartment which should be used for something else. It is difficult to site anything of a permanent nature in this area because, in order to load the torpedos, a hatch is required above the surface water-line, which is consequently located at the higher part of the pressure hull. In most submarines the diameter of the pressure hull is not much more than the length of a torpedo and hence it is not possible to have a simple vertical hatch for loading torpedos, and so a forward inclined hatch is necessary to allow weapons to be taken down at an angle. This poses a particularly difficult structural problem at the fore end of the submarine. It also causes other problems in layout because when a torpedo is being loaded into the submarine at an angle it occupies more than one deck height and so takes up most of the fore end space during the embarkation process. It may be possible to confine the disturbance to a centre line slot, which can be a passageway with removable decking for the stowage operation. In that case, compartments can be sited on either side possibly those associated with sensor processing.

On other submarines a directly opposite arrangement of the fore end from that describe above is adopted, in which the sonar is sited in the lower or chin position and the torpedo tubes in the higher position allowing more space inboard in the pressure hull in the lower part for tankage. (Figure 7.5) It could be possible with that arrangement to use the upper torpedo tubes themselves as the loading ports for weapons to be taken onboard instead of using a torpedo loading hatch. That scheme would require the bow of the submarine to be raised high enough during the torpedo embarkation process to bring the upper torpedo tubes well clear of the water. Although this arrangement overcomes some of the fore end problems it does leave unresolved the situation in which guidance wires trailing from the tubes may drift across the face of the sonar causing interference.

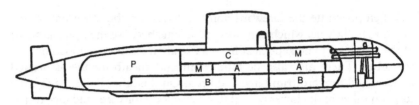

Fig. 7.5   Modern diesel/electric submarine

The lettering on Figure 7.5 and Figure 7.7 mean the following:-
    P = Propulsion and power   W = Weapon stowage
    A = Accommodation   M = Auxiliary machinery
    C = Command and control   B = Battery

A torpedo tube requires a bow cap to form a seal at the outer end of the tube in order to keep it dry for loading from inboard. There is a need for space ahead of the bow cap to allow it to swing open clear of the tube, so the hole in the outer surface of the hull at the bow has to be a little further forward of the opened cap. Although in older submarines the opening in the hull was left clear for discharge, in modern high speed boats it is highly undesirable to have openings at the bow of the vessel and therefore some form of closure is required. The closure is known as the bow shutter. Bow shutters may vary in geometry from more or less circular or rectangular if the tube is on the midships plane, to long elongated closure plates on tubes sited more towards the sides of the vessel. Side shutters also provide some guidance to the torpedo as it leaves the tube, which it needs because it has to run a considerable length with, on one side, hull structure and on the other side, water. In some circumstances this asymmetric flow condition can cause a torpedo to deviate from a straight line and that can cause damage to its tail fins on leaving the tube. The inboard end of the tube also has to have space for a closure, to keep the tube watertight when it is flooded. This closure is known as the rear door and takes a similar form to the breach loading of a large gun. It must be watertight and very secure, so some form of interlock arrangement has to be provided so that the bow cap and rear door cannot be opened at the same time. Failure to ensure that condition cannot occur would in most circumstances be catastrophic to the submarine when dived, as it would effectively breach the hull with a very large flood hole through which water could enter very rapidly. The loss of *HMS Thetis* with many of her crew in 1939 was attributed to such a failure. Once the breech lock is disengaged the force of water is capable of blowing the torpedo door open and it will then be impossible to shut. Whilst the primary safety feature is the interlock between rear door and bow cap, other safety devices are fitted. One is a test cock on the rear door to check that there is no water in the tube. (In the *Thetis* this was apparently sealed up with paint) The other is what is known as a 'Thetis clip', which is a screw-down latch on the rear door so that it cannot fly open when the breech lock is disengaged. In the event of an interlock failure even a small gap will allow water to enter, but the door can be screwed down and re-locked.

As we described in the Hydrostatics chapter there is a requirement at the fore end of a submarine for a number of special tanks associated with torpedo discharge, which are provided to enable the submarine to remain 'in trim' throughout the weapon discharge cycle. At the end of a patrol all weapons may have been discharged so that to maintain hydrostatic balance a tank volume equivalent in water to the weight of all weapons has to be arranged at the same longitudinal position as the weapons. This adds considerably to the space demands at the fore end.

It should not be overlooked that the weapons stowage compartment abaft the torpedo tubes has to be treated as a magazine because it contains not only weapons loaded with high explosives but also rather nasty and volatile propellent mixtures for their propulsion system. Great care therefore has to be exercised in the design of the compartment to minimise the risk of spillages of dangerous fuel and possible causes by which the explosive mixtures contained onboard might be set off. For that reason it is difficult to use the fore end space for other purposes and manned occupation is best avoided, though in older submarines some crew accommodation was arranged amongst the torpedoes.

To avoid some of the clashes at the forward end of the submarine other locations for the discharge of torpedoes have on occasions been adopted. These include angled tubes discharging from the sides of the pressure hull, either in the cone or dome area so that discharge occurs further aft than with bow sited arrangements. Some older submarines have one or two tubes sited aft, primarily for use if being hunted by a ship which is following them. The alternative arrangements to bow-sited tubes pose their own problems. The angled tube arrangement occupies far more space inside the pressure hull because to load the tubes, torpedoes have to be aligned across the compartment or downwards through the decks in order to get them into the tube. A hydrodynamic problem can also arise because a weapon being discharged at an angle to the flow from the side of the hull can experience a greater toppling moment causing damage to the torpedo as it leaves the tube. Toppling is less of a problem at the bow as the torpedoes are being discharged into an area close to the stagnation point where there is only a small cross flow at the tube opening.

The forward compartment of the pressure hull, which is usually the weapon stowage space, also performs a function as an escape compartment in the event of the submarine sinking in a depth of water less than its collapse depth. Escape arrangements are discussed later in Chapter 9. In the arrangement of a forward escape compartment provision has to be made for an escape hatch. Externally arrangements will be made for the docking of a rescue vehicle. The compartment must as well provide for twill trunk mass escape, lockers for escape survival suits, emergency air treatment units and built in breathing system; allowance has also to be made for the possibility of all the crew being assembled in the compartment in an emergency.

Further considerations of fore-end arrangements

7.7   If a bow sonar and forward firing torpedo tube configuration is adopted these two systems give rise to congestion in the fore end ahead of the pressure hull. However, the difficulties do not stop there.

Considerations of depth control, which we describe in a later chapter, lead to a requirement for forward hydroplanes sited well forward and preferably at hull axis level. Thus space has to be found in the fore end for the associated shafting, bearings and mechanical crank arrangement necessary to support and actuate the fore planes. In addition, the vulnerability of planes which project from the hull forward and the lesser need for them at high speed lead to a requirement to retract or house the planes, which imposes a considerable additional demand for fore end space. The operation of the planes and their disturbance to the water flow make it highly undesirable to locate them in close proximity to the sonar. Every effort has to be made to remove the planes from the vicinity of the sonar and to isolate them acoustically.

For these reasons it is sometimes decided not to fit the planes in their optimum control position and instead to move them aft and upwards. They may be sited above the pressure hull in the fore part of the casing, but this poses considerable hydrodynamic difficulties in obtaining smooth non-separated flow and in achieving flow alignment and plane effectiveness. In some submarines the planes are moved even further aft and mounted on the bridge fin. This configuration has both advantages and disadvantages in control which we discuss later.

Another cause of added congestion at the fore end are the anchoring arrangements. A submarine has occasions like any other ship when it must anchor. The normal ship anchoring arrangements are not compatible with the design of a submarine fore body. It is usual in modern submarines to provide a hydraulically powered cable winch which can be operated internally as there is little deck space to do this externally, particularly in rough weather. The anchor itself is likely to be of the mushroom type which can be drawn up into a housing underneath the bow, and so fore-end space has to be found for the anchor housing, cable fairleads, winch and cable locker external to the pressure hull.

Control room

7.8   The control room of a submarine serves the purposes of a ship control centre, operations room and, in the submerged condition, a bridge watch keeping station. This multiplicity of functions means that the space has to be arranged so that the command has an oversight of the steering and planes position, ship status boards, wireless office, sonar room, navigation systems and radar systems. At the same time the command can require access to an optical periscope position. It is the latter requirement, which remains in all submarines today, of direct optical viewing through a periscope which governs the location of the control room, as it effectively ties the position of the control room to the bridge fin arrangements outside of the hull. This can either be interpreted to require the control room to be located nearly under the bridge fin or the bridge fin to be close to the location of the control room; primarily, though, they should both be adjacent longitudinally. The optical periscope is a fixed length tube in which the upper lens must be above the water in order that the command can see the surface scene, but at the same time with the hull as far below the water as possible to avoid the possibility of collisions or giving away its position. The fixed length tube has to be capable of being lowered down so that it is enclosed within the bridge fin when not in use. Hence the scope of the periscope, i.e. its travel from raised to lowered position, has to be encompassed within the hull of the submarine. In some small submarines, it is actually allowed to penetrate through into a pressure-tight well in the keel of the submarine. In order that in the raised position the eyepiece is at the right level in the control room it will generally be necessary for the the control room to be on the uppermost deck of the submarine, though in a very large diameter boat it may be possible to lower this one deck. In other submarines an alternative approach is to have an upper conning position (an additional pressure structure) within the bridge fin. As well as the co-location of periscopes in the control room there is also a requirement for other masts and for radar and radio systems to be sited within the bridge fin area and it follows that the spaces where the data is processed and presented should be arranged around the control room. The main conning tower hatch with access to the bridge conning position will usually be located just to the fore end of the control room area.

As indicated earlier, although difficult to avoid in a single deck submarine, it is undesirable that the control room should be on a through passage between other areas of the submarine. Ideally, the control room area should be sited in effectively a cul-de-sac, where only those crew members involved in changing watch need enter and leave the space. This consideration points to the idea of locating the control room in the forward compartment of multi-deck arrangements above the weapon stowage compartment. That disposition has the advantage of siting the control room in a cul-de-sac in terms of personnel movement. Although it poses problems of weapon stowage, they might be solved with an open control room which had space on the mid-line for weapon handling. The major current disadvantage is that such a forward position is incompatible with the location of the bridge fin for periscopes to be physically sited in the control room. On the other hand it would remove from the control room the 'wet' area usually associated with the conning tower hatch. It may be possible to contrive an arrangement where the periscopes are towards the front of the bridge fin and at the aftermost part of a forward control room. Another solution (but no easy task) would be to persuade the submarine command that they do not need physical/optical access through the periscope of the surface; by means of modern technology it is possible to use a television camera to give an electronically produced picture/display of the above water situation using a remotely sited camera sensor on a retractable mast.

7.9 In addition to the assembly of masts and periscopes required for the operation of the submarine, it is usual to arrange for the ventilation mast or snort mast also to be housed within the fairing of the bridge fin. In the snorting arrangements of earlier submarines piped air supply was taken directly into the engine room, while the exhaust air pipe was re-directed back and up to the rear of the bridge fin, to exhaust below but near the surface. Ventilation of the rest of the boat was achieved in a somewhat haphazard fashion, either by the diesel engines drawing some air to be replaced by fresh air from the fore end of the boat or by circulating air from and to the engine room. On modern submarines the object of the snort head valve and mast is to act as a means of ventilation for the complete boat. Air is taken down into the lower part of the fore end of the boat and, as the usual practice when snorting is to be charging batteries, air is drawn across the top of the battery compartments to clear the hydrogen being evolved in the charging process. The air is then drawn back through the other compartments in the boat to the engine room. Since the snort head valve is only slightly above the water surface and from time to time will dip below it, a certain amount of water is brought in down the masts. Much of the water can be separated out by means of by a centrifuge arrangement before it enters the hull, but a certain amount of water will be carried down into the boat. To cater for this water carry-over, provision is made for some tank space into which the ventilation mast discharges its air/water mixture, and there the water is separated out and the air dried before being drawn over the batteries which do not tolerate salt water. Where the snort mast enters the hull there is a large shut-off valve to preserve the safety of the submarine when dived, when the mast is flooded. Though the snort head valve is normally designed to shut automatically when dipped below the waves, there is a danger of it sticking open or in some way failing to operate; as a precaution the piping system is fitted with an emergency rapidly shutting valve which normally takes the form of a flap capable of being driven shut by the heavy ingress of water whilst remaining open to air flow.

Accommodation for Crew

7.10   The accommodation for the crew in a multi-deck submarine is most conveniently arranged within the mid-body of the hull, between the weapons and control areas forward and the propulsion plant aft. In large submarines the accommodation can be laid out as a coherent block on one or two decks. In smaller boats accommodation has to be contrived in such spaces as are left after operational considerations have fixed major locations of equipment. It has to be kept in mind that the crew may have to remain within the enclosed environment of a dived submarine for considerable periods and even on the surface access to the bridge or casing is severely restricted. Thus, in design, the aim is to provide living quarters of good standard for sleeping, eating, recreation and domestic services within the limited space available. In most submarines provision rooms and chilled spaces are provided for the stowage of fresh food to be prepared on board for eating. These facilities take up a substantial amount of space on a long patrol vessel and also give rise to a waste disposal problem. Some space saving can be made by arranging for pre-prepared fresh food ready for cooking or even airline type prepared meals requiring only heating.

Though submariners adapt to the communal living conditions, some opportunity for individual privacy should be afforded in the arrangements, even if no more than a curtained-off bunk space.

Bathroom and toilet facilities also have to be provided, and these can usually be segregated on reasonable sized submarines. Personal hygiene is important in such confined conditions, but the provision of potable water may be limited as it involves stored energy consumption, and so salt water may have to be used for purposes where potable water is not essential. In the case of these facilities there is difficulty in arranging ready access to bathrooms from living spaces, whilst keeping water and odours as far away as possible.

The systems and tanks involved in waste disposal are described in a later chapter.

Battery compartment

7.11   For diesel electric submarines the batteries occupy considerable space and, because of their weight, are installed in the lower part of the hull. There will be at least two battery compartments, each with about 200 individual batteries. These will be stacked close together so that each cell can be readily linked to its neighbours to provide a high voltage section. Despite the close stacking there is scope for variation in height between rows. Thus along the compartment the outermost rows of batteries are raised to follow, in step fashion, the circular shape of the lower hull. (Figure 7.6) Some clearance from the hull is required at the bottom to reduce the possibility of damage to the battery cells under shock. The stepped arrangement is mirrored at the top of the battery compartment, providing a useful space under the overhead deck for access and ventilation of the space. Because of the possibility of acid spills, the compartment must be lined with a thick impermeable rubber coating. Seawater is a menace to batteries, with the possibility of generating chlorine gas, and to prevent that hazard the top of the compartment and the access-removal hatches are sealed to prevent water leakage. If at all possible, battery compartments should be kept clear of areas of the submarine where water spills may occur.

Propulsion plant compartments

7.12   Moving aft in the internal arrangements, the rear half of the submarine is almost entirely occupied by the propulsion plant. In nuclear powered vessels the reactor with its ancillary machinery and shield occupy the whole of the centre block of the hull. The sheer density of the reactor plant requires it to be close to the longitudinal centre of gravity, as otherwise there could be a balance problem. The reactor plant can also be a determinant of the maximum diameter of the hull. Safety and radiation hazards limit access to the Reactor Compartment and access to the after ends from forward requires a shielded tunnel through the space. It is appropriate for at least one of the reactor compartment bulkheads to be a major pressure hull structure as it will also have to support and distribute some of the concentrated weight. With this effective block to fore and aft access it is usual to treat the spaces forward separately from the after end spaces and locate auxiliary machinery and air treatment units in a space just ahead of the Reactor Compartment.

Aft of the Reactor Compartment the dominant feature is the space for main electrical generating plant, turbo-generators and emergency diesel generators, together with the Main Distribution Switch Boards. The upper part of the hull may be a site for the Machinery and Reactor Control Centre and the lower part occupied partly by the turbo-generator condensers but there is usually space for some tankage, for compensating, lub oil, fuel and fresh water.

Further aft the narrowing hull is almost completely occupied by the main propulsion turbines with their condensers and associated piping

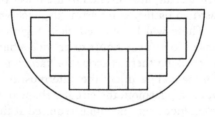

Fig. 7.6   Battery stowage

and auxiliary systems. The condensers require very large sea-water pipes and hull valves. The arrangement of this space depends on the turbine/gearbox configuration and its alignment with the propeller shaft.

Right at the after end will be a shaft mounted electric motor, main actuator rams for rudders and hydroplanes and the after trim tank. This compartment may also form the alternative escape compartment in the event of flooding forward.

7.13   Similar but different arrangements apply to a conventional submarine. Again the propulsion plant occupies almost half the hull. The main diesel generators usually occupy the first compartment aft of amidships, which is a relatively dense compartment needing to be close to mid-length. The diesels also require to be fairly close to the rear of the bridge fin to reduce the run of exhaust and induction piping to the snort mast.

If, as is usually the case, direct drive is not catered for as a propulsion option the diesels and their electrical generators are not required to align to the propulsion shafting, which gives some freedom in arrangement both vertically and horizontally. It should be remembered, however, that if the diesel generators were sited at the sides of the hull there would be less available headroom due to the overhead curve of the hull. Whilst many submarines now make provision for a large section of the hull to be removed to conduct major overhauls on engines, many maintenance tasks will still be conducted in enclosed conditions with a relatively small access hatch for men and equipment. Thus, at the detail design stage, attention has to be given to access and readiness of removal of many parts of the machinery which may require repair or replacement.

Aft of the main diesel generator compartment will be the main electrical switchboard. Though the electrical distribution demands of a conventional submarine are necessarily less than those of a nuclear vessel there is nevertheless a major requirement for the switching of the high voltage direct current associated with the electrical propulsion system.

The aftermost compartment is dominated by the propulsion motors, control surface actuators and after trim tank, and so can be a very congested space particularly if after torpedo tubes are fitted (though stern tubes have rarely been adopted in modern submarine designs).

Aft of the pressure hull the outer form is narrowing rapidly towards the propulsor. Within this space, however, some form of after MBT has to be accommodated. Furthermore, the main drive shaft has to be taken through from the pressure hull shaft seal to the propulsor hub. There is also a requirement for rudders and hydroplanes at the tail. Because the shafts of the control surfaces would otherwise intersect the propulsor shaft line at right angles, they have to be split and displaced to either side of the drive shaft by massive yoke arrangements. The yokes also act as the levers for actuation of the control surfaces by push rods from internally stowed hydraulic rams. It will be appreciated that these multiple demands on a very cramped space necessitate careful detailed design to avoid interference and achieve workable and maintainable arrangements.

REVIEW

7.14 It will be realised from this chapter that while adequate volume/space provision may have been made at the concept stage of sizing the submarine, the utilisation of that space requires meticulous detailed study. It is possible that, though in broad terms sufficient space has been allowed, the practical utilisation of space cannot be fully realised. That possibility is only too likely to arise at the fore and after ends of the vessel. It is advisable in consequence that layouts should be started very early in the design process because they may necessitate an increase in the size of the vessel, with all the upheaval consequences which then ensue, and the sooner the need is identified the better. The general arrangements of several submarine designs shows some of the ways in which this architecture has been solved. (Figure 7.7)

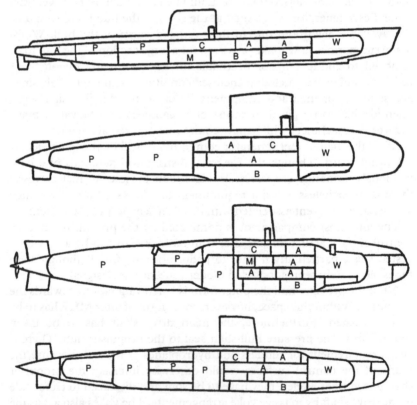

Fig. 7.7   Typical submarine arrangements

8 DYNAMICS AND CONTROL

INTRODUCTION

8.1 Historically, it was not until the advent of the nuclear submarine with its capability for sustained high speed that the focus of attention of the submarine designer moved from the provision of adequate means for control of motion in the vertical plane – particularly at periscope depth – to achievement also of an appropriate balance between manoeuvrability and dynamic stability, with the emphasis again on motion in the vertical plane. Concentration of attention on depth changing and keeping rather than on course changing and keeping was natural because of the inherent risks in the high speed submarine – even with better diving depth capability through increased pressure hull strength – of accidentally exceeding the allowed maximum depth, a hazard regarded as potentially more dangerous than inadvertent surfacing.

As can be appreciated, there are similarities between the submarine manoeuvring submerged and the airship – though there are also significant differences – and in fact early theoretical and experimental investigations into submarine dynamic stability and control, in the 1940s and 1950s, initially drew on corresponding investigations for airships in the 1920s and 1930s. Subsequent research, specific to submarines, has provided a body of knowledge of which it behoves the submarine designer to have a basic understanding, even though dynamics and control are not primary considerations in determining the size and shape of the submarine at the concept stage.

It might seem from the latter observation that, once size and form have been determined by the dominant considerations, provision of the means for achieving the desired control characteristics could be left to specialist hydrodynamicists and control engineers for development subsequent to the concept stage. While that is to some extent so, it is nevertheless important for the designer in configuring the submarine to appreciate how that configuration influences the control characteristics, particularly if the operational requirements were to emphasise aspects such as high agility, or very stable platform, or accurate positioning.

By way of illustration of the design interactions, the general shape of the external form of a submarine, e.g. short and fat or long and thin (Figure 8.1), confers hydrodynamic characteristics and also limits the lever arms of control surfaces relative to the centre of gravity of the vessel; the choice of location and space for rudder and hydroplanes determines their capability to contribute to depth keeping and changing; and the location and size of major appendages, such as bridge fin, casing, and form of bow and stern, all have a significant effect on the way the submarine behaves when manoeuvring.

The designer must in consequence be sufficiently aware of all these issues and what he can do to favourably influence the outcome, and that is our objective in this chapter.

Although in what follows we refer to rudder in the singular, it is often the practice to have upper and lower rudders.

SOME BASIC CONCEPTS

Freedom of motion

8.2 A submerged submarine has freedom to move in all directions that constitute the six degrees of freedom which, in naval architecture, are termed surge, sway and heave for the bodily translations along the three axes of the vessel (namely, longitudinal, athwartships and vertical) and are termed roll, pitch and yaw for the angular rotations about those axes. Although there is usually some interaction between the motions, it is often a sufficient simplification to treat them in uncoupled groups. (Figure 8.2) When that applies, surge (the change of speed of the submarine in the direction of its longitudinal axis) is treated as a single, independent motion related to the powering and resistance of the vessel; the

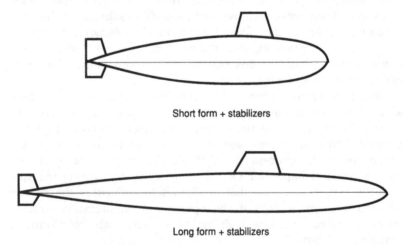

Short form + stabilizers

Long form + stabilizers

Fig. 8.1   Effect of L/D ratio

motions in the horizontal plane, i.e. sway (the sideways movement) and yaw (the rotation in heading) are treated as a coupled pair; the motions in the vertical plane, i.e. heave (the up and down movement) and pitch (the angular attitude) are also treated as a coupled pair; and roll (the rotation about the longitudinal axis) is treated as single, independent motion, even though it is closely related to the turning motion.

Motion control

Looking at the motions in the coupled groups described above, the customary approaches to their control are as follows:

(a) Surge: This motion is the outcome of the variation between two longitudinal forces, the thrust from the propulsor and the resistance of the vessel to forward motion. When the submarine is proceeding, sufficiently deep, on a level path at constant speed, the forces are equal and opposite, but if it changes course and/or depth the other motions will alter the resistance and a speed variation will result. Control of surge is not usually attempted, but could be effected by changing the propulsor RPM.

(b) Yaw and Sway: The means of control of this coupled pair is by rudders at the after end of the submarine. Rudder operation effects control of heading or rate of turn by causing the vessel to take up an angle of yaw; sway is a consequence of yaw and generally no attempt is made to control it directly.

(c) Pitch and Heave: It is the freedom of a submerged submarine to move in the vertical plane that differentiates it

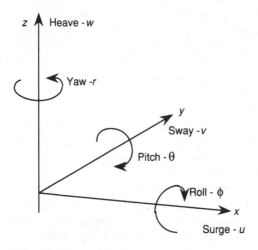

Fig. 8.2   Freedom of motion

from the surface ship, for which the pitch and heave motions are relatively small and determined by surface waves and the hydrostatics of the ship's waterplane. In submarines, the means of control of this coupled pair is usually by two sets of control surfaces known as the hydroplanes, one set forward and one aft. With that approach, it is possible to control pitch and heave independently. In older, slow speed, submarines it was common practice, in manual control, to have two planesmen, one to each set of hydroplanes, with the forward planesman controlling the depth of the submarine and the after planesman controlling pitch angle. As we shall see later in the chapter there is sound logic in the practice at slow speeds, though at higher speeds the need for separation of control in that way diminishes and then coupled control can be taken over by the after hydroplanes alone, the forward hydroplanes being zeroed.

(d) Roll: It is not usual to provide control of roll in submarines, unlike many surface ships in which roll stabiliser fins are employed. In submarines, any asymmetric moment tending to cause roll is countered by the hydrostatic restoring moment due to the centre of gravity being below the centre of buoyancy. As we shall see a transient rolling moment is caused when a submarine turns at speed submerged, which can be very disadvantageous, and so ways of reducing the effect are worth pursuing.

OPERATIONAL REQUIREMENTS

8.3   To provide a background to consideration of the dynamics of submarine operation submerged, we first discuss the requirements for control of the motions of the vessel. These are rarely expressed explicitly in statements of requirements or specifications, largely because they are taken for granted, but are nevertheless necessary to understanding of the purposes sought in the area of dynamic stability and control.

Speed Aspects

The achievement and maintenance of forward speed are mainly the province of the designers of the propulsion plant, transmission systems and propulsor, and call for attention to the whole speed range from virtually stopped to full power. When the submarine manoeuvres in depth and course the additional resistance caused by the lateral motions will result in a reduction in speed. For example, in a sharp turn the speed on the turn will be considerably lower than on a straight course at the same power levels, which is undesirable as the rate of change of heading may actually be lower than in a less tight turn. In other circumstances, the ability to accelerate rapidly from slow speed can be operationally advantageous both for evasive purposes and for safety reasons. Safety is a consideration we touch on later under the heading of emergency recovery, which deals with hazards like flooding when travelling slowly and jammed after hydroplanes when travelling at speed. In the former situation, recovery action would include accelerating to generate more hydrodynamic forces, and in the latter situation recovery action would include slowing down to reduce hydrodynamic forces. Either way, there is a connection between propulsion and control considerations which needs to be kept in mind.

Course aspects

For motion in the horizontal plane, the basic need is to maintain a set course with little variation in heading, for which purpose the submarine should have adequate dynamic stability so that course-keeping does not require continuous activity of the rudder. However, there will be other circumstances in which the vessel needs to be able to turn rapidly off course in the process of carrying out an evasive manoeuvre. As already discussed, to make a very tight turn might not lead to the fastest rate of change of heading, and so a less tight turn might be better in causing less loss of speed on the turn. In design, there is in consequence, not only the issue of compromise between good course keeping and changing but also that of how to achieve good agility for evasive purposes.

Depth aspects

There are similar requirements for motion in the vertical plane, the simplest of which is the ability to maintain steady depth without undue activity of the hydroplanes. While relatively deep there is little in the way of varying external force acting on the submarine and provision of adequate dynamic stability suffices for that purpose. As the submarine approaches the water surface it may encounter the effects of wave action which generate disturbing forces tending to cause it to heave and pitch; there will also be suction forces on the hull due to its proximity to the surface while underway. Although the submarine's dynamic stability influences its response to wave excitation, depth control can only be achieved by use of the hydroplanes, i.e. the 'stick fixed' mode is no longer what matters but whether sufficient control forces can be exercised to counter the cyclic forces due to action of the waves and the more steady suction forces.

As well as being able to maintain depth in these different circumstances, the submarine needs to have the capability to change depth in a controlled fashion, sometimes at quite a high rate for evasive purposes, calling for agility in the vertical as well as the horizontal plane. The usual way in which a depth changing manoeuvre is carried out is to pitch the boat up or down as appropriate to the direction in which it is required to change depth and then to drive it at that angle. Rate of change of depth is then determined by the product of the pitch angle and the speed achieved at that angle. It is the practice to limit the maximum pitch angle at which depth changing is effected in this way to about 20°, as it is difficult for the crew to work effectively at larger angles and such a large inclination might also cause malfunction of machinery and equipments on board. At high speed the submarine can transit from near the surface to maximum operating depth – a distance of just a few times its length (Figure 8.3) – in a matter of a minute or so, which leaves little time for recovery action in the event of a control system failure. Thus the maximum pitch limit is rarely used and a more usual pitch angle would be 5-10°.

A quite different type of depth changing operation is that in which the submarine requires to change depth slowly while maintaining horizontal attitude. That situation can arise when approaching periscope depth when it is undesirable to show too much of a mast above the water surface. Under-ice operation calls for rather similar precision of control of depth and attitude. As we shall see later, slow speed operation of the submarine involves substantially different considerations from high speed operation as regards depth keeping and changing, because of the increasing influence at slow speed of hydrostatic forces and moments in relation

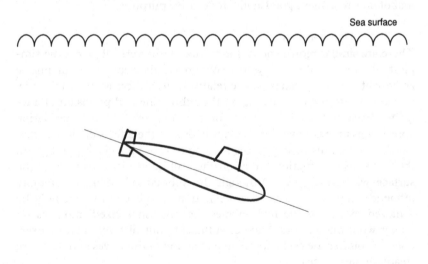

Fig. 8.3   Operating scope

to hydrodynamic forces and moments. From the designer's point of view it is important to have a feel in designing the submarine for depth changing and keeping, and how a balance can be struck between dynamic stability and agility and safety while exploiting to the full the freedom that manoeuvring in depth confers on the modern submarine.

Roll aspects

It is not usual to specify any requirements for control of the roll motion, though the designer needs to know the causes of the phenomenon termed 'snap roll', which we describe later, and what steps can be taken to minimise it. It is the practice to lay down a limit on maximum roll angle – 50° is not untypical – to which equipments and machinery on board the submarine have to be designed. Whilst this is set as a design limit it is not acceptable for normal operation.

In addition to the foregoing general control requirements, some submarines may have to meet particular requirements for track-keeping, in both the vertical and horizontal planes, associated with sensor operations, e.g. when conducting sonar search or oceanographic surveying. Yet again, some vessels may have a more specific requirement for control of attitude and/or position, e.g. a submarine rescue vessel needs to be able to align itself and mate with the superstructure casing of a stricken submarine in way of escape hatches. Similar requirements arise with Remote Operated Vehicles, (ROVs), operating at seabed platforms used for offshore oil and gas production, where there may be a need to take up a precise attitude to carry out a maintenance task. We do not treat the design problems which arise with those highly specialised underwater vessels, though the theory now addressed would be applicable to such problems.

EQUATIONS OF MOTION OF A SUBMARINE

Conventions

8.4   Although it is not our intention to go very far into the theory of submarine dynamics, it is desirable for a general understanding to provide a rudimentary treatment and this involves some preliminary discussion of the form of the equations of motion of a submarine. We do not go into their derivation, which can be found in a suitable textbook. To that end it is necessary to define the axes and coordinate system employed for the purpose and these are illustrated in Figure 8.2. The set of axes commonly used is aligned to the longitudinal, vertical and athwartships geometry of the submarine, assumed to be moving in three dimensions in hydrospace, the centre of the set being either at the geometric centre of the boat or, more conveniently, at its centre of gravity. This choice of axis system overcomes some of the problems of coupling between the various motion components, thought it does lead to other complications as regards forces such as gravity which are related to a spatial set of axes.

It will be recognised how the equations that follow fall into the pattern of the coupled motion groups described earlier.

The surge equation

$$m\dot{u} = X_P + X_U + X_M + X_A$$

This shows that the rigid body mass of the submarine times its acceleration in the direction of its longitudinal axis is equal to the sum of the forces acting on it in that direction. These forces comprise: the propulsor thrust; the hydrodynamic resistance appropriate to its motion in the direction of its longitudinal axis; the summation of additional drag forces arising from any lateral motion which might be occurring; and the hydrodynamic forces associated with accelerated motion in the direction of its longitudinal axis, commonly known as the 'added mass' term.

Horizontal plane equations

$$m(\dot{v} + rU) = Y_V + Y_{\dot{V}} + Y_R + Y_{Cont}$$

and

$$I_{zz}\dot{r} = N_V + N_R + N_{\dot{R}} + N_{Cont}$$

The first of these equations relates the product of the rigid body mass and its sideways acceleration to the summation of the hydrodynamic forces acting on the hull in the sideways direction. The second equation relates the product of the rotary inertia of this rigid body about a vertical axis through the centre of gravity and its angular acceleration in yaw to the summation of the horizontal moments of the hydrodynamic forces acting on the hull.

Vertical plane equations

$$m(\dot{w} - qU) = Z_W + Z_\theta + Z_{\dot{W}} + Z_{Cont}$$

and

$$I_{yy}\dot{q} = M_W + M_Q + M_\theta + M_{CONT}$$

The first of these equations relates the product of the rigid body mass and its acceleration in the vertical direction to the summation of the hydrodynamic forces acting on the hull in that direction. The second equation relates the product of the rotary inertia of the rigid body about a horizontal axis through the centre of gravity and its angular acceleration in pitch to the summation of the vertical moments of the hydrodynamic forces acting on the hull, augmented in this case by the hydrostatic restoring moment due to the departure of the axis of the submarine from the horizontal.

Roll equation

$$I_{XX}\phi = K_V + K_R + K_\phi$$

This relates the product of the rotary inertia of the rigid body about its longitudinal axis and its angular acceleration in roll to the summation of the athwartships moments acting on the hull due to the hydrodynamic forces arising from the other motions, augmented in this case by the hydrostatic restoring moment due to the departure of the submarine from the vertical.

Control forces

The hydrodynamic forces which are taken into account in the foregoing equations are separated into those regarded as body forces and those regarded as control forces. The convention adopted is that the forces generated by the motion of the body include the contributions of _all_ appendages, not only any stabiliser fins but also the control surfaces themselves assumed to be at their zero settings, i.e. in the 'stick fixed' mode; control forces with this convention are those forces which are only brought into effect when the control surfaces are deflected from their zero settings.

HYDRODYNAMIC DERIVATIVES

8.5 To proceed with analysis of the above equations of motion, it is necessary to describe the hydrodynamic forces and moments in suitable terms, and the widely used approach – originally introduced in the aerodynamic field – is the adoption of 'derivatives'. Essentially the assumption is made that if the motions of the body can be taken to be reasonably small departures from an initial straight line motion, the hydrodynamic forces and moments acting in consequence on the body can be taken to be directly proportional to the associated small changes in velocity. This relationship is represented by coefficients known as hydrodynamic derivatives. Thus, for example, if there is a small sway velocity v, the hydrodynamic force Y arising from that motion is expressed as $Y = Y_v v$, where Y_v is the linear force derivative in respect of sway velocity.

The adoption of the linear derivatives procedure is a helpful way in which to examine some of the basic characteristics of stability and control in submarines, but it is important to keep in mind that there are many circumstances in which the motions will not conform with the assumption of linearity entailed in the approach. In order to model the behaviour of a submarine undertaking such extreme manoeuvres that linearity can no longer be assumed, it is necessary to take into account higher order terms in the equations of motion. While this is feasible, analysis becomes much more complicated and it is difficult to draw the general conclusions which linear analysis allows.

Even for modest manoeuvres, linear representation of the surge forces is unsatisfactory and to obtain meaningful results it is necessary to adopt second order terms where the velocity is squared in order to achieve symmetry in the drag terms associated with a turn to port or to starboard. This aspect also occurs in the aerodynamics field, where the additional drag due to incidence on a lifting aerofoil is expressed in terms of the square of the lift coefficient.

It should be noted that the derivative form expressed is dimensional. In much of the literature a form of the equations is used in non-dimensional terms denoted by a prime. The relationship is such that, for example:-

$$Y'_v = Y_V / \frac{1}{2}\rho L^2 \, U$$

where L is usually taken as the length of the vessel. The non-dimensional form of derivative is usually assumed to be a constant characteristic of the geometry of the body. Hence in the dimensional form used in this brief exposition, the dimensional hydrodynamic derivatives are dependent on the square of the speed. This will be seen to be significant where there is an interplay between hydrostatic and hydrodynamic forces and moments.

STABILITY AND CONTROL IN THE HORIZONTAL PLANE

Equations in derivative form

8.6 Applying the derivative approach to the equations in Section 8.4, and also adapting them to include control terms appropriate to the deflection of the rudder to angle δ, the linearised equations in sway velocity v and yaw rate r become:

$$m(\dot{v} + rU) = Y_v v + Y_r r + Y_{\dot{v}}\dot{v} + Y_\delta \delta$$

and

$$I_{zz}\dot{r} = N_v v + N_r r + N_{\dot{r}}\dot{r} + N_\delta \delta$$

Thus we have two equations for two unknown parameters v and r and with an input parameter, the control term δ.

Dynamic stability

We start with consideration of the behaviour of a submarine initially travelling on a straight course with no control input, i.e. $\delta = 0$. The equations can be re-arranged in the simplified form

$$(m - Y_{\dot{v}})\, \dot{v} \;=\; Y_v v \;+\; (Y_r - mU)\, r$$

and

$$(I_{zz} - N_{\dot{r}})\, \dot{r} \;=\; N_v v \;+\; N_r r$$

These coupled differential equations in the variables v and r can be readily solved by means of transforms to a single equation for either v or r, which is a second order linear homogeneous equation. The roots of this equation can be derived so as to indicate the dynamic stability of the submarine moving in the way assumed. The general solution of the equation is in the form of exponential terms in which the exponent can in principle be positive, negative or imaginary. (Figure 8.4) To have complex exponents would mean an oscillatory response of the submarine to a transient disturbance, but this case never arises in practice, and so the

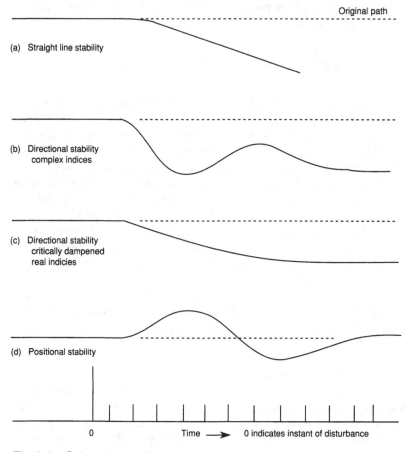

(a) Straight line stability

(b) Directional stability
     complex indices

(c) Directional stability
     critically dampened
     real indicies

(d) Positional stability

0   Time ⟶   0 indicates instant of disturbance

Fig. 8.4   Submarine stability modes

disturbed behaviour is characterised by real exponents. If those were positive, any small disturbance from the initial straight path would increase with time and so the submarine would increasingly diverge from its original course unless control action were taken. For the submarine to have good course-keeping capability, the real exponents should be negative so that the response to small disturbances is damped out even in the absence of any control action.

The usual way in which to assess the stability implications of the single equation is to apply the Routh stability criteria. In the present case, these require that the constant term, i.e. the third term of the equation, should be positive. The term is a combination of the force and moment derivatives for the sway and yaw motions, and the determinant of dynamic stability is:-

$$N_r Y_v - N_v (Y_r - mU) > 0$$

By dividing through by the force terms Y_v and $(Y_r\text{-}mU)$ this condition can be expressed as:

$$\frac{N_r}{(Y_r - mU)} - \frac{N_v}{Y_v} > 0$$

Expressed in that way, it can be seen that the first term is the ratio of the moment due to the rotational motion to the force due to the rotational motion, and so can be considered as a measure of the point of action of that force, represented by say $\bar{x}_r$. Similarly, the second term is the ratio of the moment due to the sway motion to the force due to the sway motion and so can be considered as a measure of the point of action of that force, represented by say $\bar{x}_v$. It follows that the criteria for dynamic stability of disturbed motion in the horizontal plane is that: $(\bar{x}_r - \bar{x}_v) > 0$, i.e. that the force due to the rotational motion component should act at a point further forward than the force due to the sway motion component. This makes sense because the sway force acts to increase the transient deviation off course while the rotational force acts to decrease it. As we observe later an unappended streamlined body is unable to satisfy this requirement, and it is necessary for the purpose to add fin area aft – which can include the rudder – which serves both to reduce $\bar{x}_v$ and to increase $\bar{x}_r$.

Steered motion

Having established how a submarine can be made dynamically stable for motion in the horizontal plane, we go on to examine the control effectiveness of the rudder in course-keeping and changing. For that purpose we return to the equations of motion including the δ forces generated by the rudder.

These equations can be solved to provide the full linear solutions for sway velocity and yaw rate due to the rudder deflection δ. However, because our interest is in control effectiveness, it is simpler to disregard transients and evaluate just the steady state motion, that is, to take the submarine to be turning steadily under rudder action, omitting all higher order derivative terms and retaining just the constant velocity terms. In that way we are left with the following pair of equations:

$$Y_v v + (Y_r - mU)\, r + Y_\delta \delta = 0$$

and

$$N_v v + N_r r + N_\delta \delta = 0$$

These can be solved to give the steady rate of turn r as a function of the rudder deflection δ, namely:

$$\frac{r}{\delta} = \frac{Y_\delta}{(Y_r - mU)}\,\frac{(x_v - x_\delta)}{(x_r - x_v)}$$

This relationship shows that the best way to achieve control effectiveness is to arrange for the point of action of the rudder force and the point of action of the resultant of the sway hydrodynamic forces on the hull of the submarine to be as far apart as possible, which means locating the rudder as far aft as can be contrived. For the same reason, a forward location for a rudder is not good for control effectiveness. (This, it will be recognised, leaves a question to be answered: why have forward hydroplanes for motion in the vertical plane? We deal with that question later).

It is also to be noted that the denominator contains the stability term $(\bar{x}_r - \bar{x}_v)$. Thus a very stable vessel will be less responsive to the rudder than a marginally stable configuration.

Although the above equation for r/δ indicates that best control effectiveness can be obtained by making Y_δ large, it should be recognised that to do so could lead to undue rudder sensitivity because quite small rudder deflections would lead to relatively large control forces, and so the helmsman in trying to steer a straight path would be constrained to using very small movements of the rudder. If those movements were within or not much more that the backlash of the steering gear it would be difficult for the helmsman to maintain a steady course even if the submarine were dynamically stable. In consequence, the designer should aim for a balance between the rudder effectiveness as represented by Y_δ and the maximum rudder angle setting, bearing in mind that for accurate course control the rudder should be allowed reasonable amounts of deflection.

Turning now to course changing, in general, from lifting surface theory, it is to be expected that stall on a foil will occur at an angle of deflection of around 10°. Though a rudder may be deflected by an angle of

35°, once the turn is fully developed, the swing outwards of the stern will reduce the effective angle of attack. Thus the rudder is unlikely to reach a stall angle except in the transient. In some cases the inflow angle will result in a reversed angle of attack at the rudder. The rudder force would then act against tighter turning and a sharper turn would be possible if the rudder angle could be increased, say, to 45°. (Figure 8.5)

It follows from this brief review of the factors involved in course keeping and changing that from the designer's point of view there are issues to be resolved about the rudder and steering gear, such as what are appropriate choices for Y_δ, for maximum value of δ and for maximum rate of putting the rudder over, and these we return to later in the chapter.

STABILITY AND CONTROL IN THE VERTICAL PLANE

Equations in derivative form

8.7   Here we are concerned with vertical motion of the submarine at velocity w and the pitch motion in terms of the pitch angle θ and rate of change of pitch angle θ. Also we are concerned with two sets of control surfaces and so two deflection angles, δ_f of the forward hydroplanes and d_a of the after hydroplanes. Applying the derivative approach to the equations we get:

$$m\,(\dot{w} \, - \, qU) \; = \; Z_w w \; + \; Z_q q \; + \; Z_{\delta f}\delta_f \; + \; Z_{\delta a}\delta_a$$

and

$$I_{yy}\dot{q} \; = \; M_w w \; + \; M_q q \; + \; \mathrm{M}_\theta \theta \; + \; M_{\delta f}\delta_f \; + \; M_{\delta a}\delta_a$$

$$(M_\theta \; = \; W \cdot \overline{BG})$$

It will be noted that the hydrostatic restoring moment term in the second equation is related to the actual pitch angle, which indicates that in the

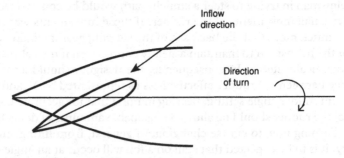

Fig. 8.5   Reverse force on rudder

vertical plane – unlike the horizontal plane – there _is_ a preferential direction, namely, a level path. As a consequence, the two equations not only contain the dependent variables w and q, but also θ, and two input parameters δ_f and δ_a.

Dynamic stability

To start with, we consider the behaviour of the submarine initially travelling along a straight and level path without any control input, i.e. with $\delta_f = \delta_a = 0$.

The equations can be re-arranged in the simplified form:

$$m(\dot{w} - qU) = Z_w w + Z_q q$$

and

$$I_{yy}\dot{q} = M_w w + M_w q + W \cdot \overline{BG}\,\theta$$

The solution is a third order differential equation, which means that there will be three roots in the solution; also the second equation is speed dependent because, whereas the hydrodynamic terms vary in magnitude with speed squared, the hydrostatic moment term is independent of speed. It follows that the roots of the equation will depend on the speed at which a solution is sought.

At sufficiently high speeds, the hydrostatic term will be much smaller than the hydrodynamic terms. If it is neglected, the second equation reverts to a second order differential equation similar in form to that derived for unsteered motion in the horizontal plane. By pursuing the approach described earlier, involving application of the Routh stability criteria, it can be shown that a similar requirement exists for dynamic stability of disturbed motion in the vertical plane to that in the horizontal plane, namely, that the force due to the rotational component of the motion should act at a point further forward than that of the force due to the heaving motion component. However, although the analogy with the horizontal plane is close at high speeds, the disturbed motion in the vertical plane will still differ in the retention of a vestigial preference for a level path.

At slower speeds, when the hydrostatic term is not negligible, the third order differential equation which then applies allows the possibility of complex roots. In fact, the general form of the solution to the equation corresponds to there being a damped oscillatory characteristic to the disturbed motion in the absence of control forces. The $\overline{BG}$ term is effectively a spring constant in the equation and so governs the frequency of the oscillation and the question for the designer is how much damping should be provided: too little could result in large oscillatory excursions after a transient disturbance and too much could result in a sluggish response.

When the possibility of achieving high speeds submerged first came into prospect, the inclination of some designers was to err on what seemed to be the safe side and to provide considerable stabilising fin area aft, accepting the adverse impact on manoeuvrability in the vertical plane. However, two considerations came to be recognised as making that policy unattractive: the first was that at slower speeds the hydrostatic moment acts to augment the hydrodynamic stabilising effect, causing the submarine to be very sluggish indeed in the vertical plane; the second was that it could be argued that the real virtue in the vertical plane is to be able to level out after a disturbance, whereas if the initial path of the submarine is inclined, whether up or down, excessive dynamic stability would cause it to try to remain on that path, which could be undesirable.

In consequence today, for submarines capable of attaining high speeds, the option usually taken up is to aim for slightly less than what is termed 'critical' damping (a concept employed in the study of mechanical vibrations which relates to avoidance of overshoot after a transient disturbance) (Figure 8.6). The means of achieving that outcome is to

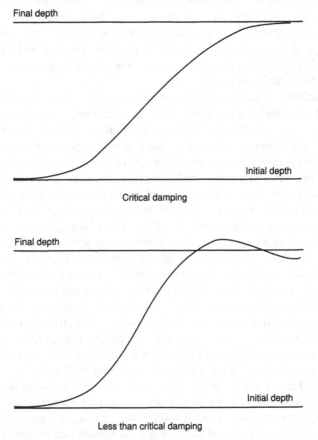

Fig. 8.6   Depth changing

add fin area aft – which can include the after hydroplanes because, as previously observed, dynamic stability is related to the 'stick-fixed' condition of the control surfaces. It will be appreciated that forward hydroplanes in those circumstances have a destabilising effect, but, as they are smaller than the after hydroplanes, the latter – augmented by the after fins if fitted – dominate the outcome.

Motion with control surfaces operating:

8.8   To start with, we will consider the equations of motion in derivative form in the particular circumstances in which the submarine proceeds on an inclined path at a steady pitch angle, i.e. at a constant vertical velocity. The relevant equations are:

$$Z_\alpha \alpha + Z_{\delta c} \delta_c = 0 \qquad Z_{\delta c} \delta c \text{ is the combined effect of}$$

$$\text{control forces i.e.} = Z_{\delta f} \delta_f + Z_{\delta a} \delta_a$$

and $\qquad M_\alpha \alpha + M_\theta \theta + M_{\delta c} \delta_c = 0 \quad \alpha = \text{vertical drift angle} = \dfrac{W}{U}$

From the first we can derive the vertical velocity component as a function of the control surface angles of deflection to give:

$$\alpha = - \frac{Z_{\delta c} \delta_c}{Z_\alpha}$$

This shows that the magnitude of the vertical velocity is a measure of the relationship between the control force and the hydrodynamic resistance of the submarine to bodily up or down motion.

From the second equation, using the solution to the first, we can derive the pitch angle as a function of the control surfaces angles of deflection to give:

$$\theta = \frac{Z_{\delta c} \delta_c \, (x_\alpha - x_c)}{M_\theta}$$

where $x_c = \dfrac{M \delta_c}{Z_{\delta c}}$ = effective location of control force and

$x_\alpha = \dfrac{M_\alpha}{Z_\alpha}$ = effective location of heave force.

The term in brackets depicts the relationship between the location of the resultant control force and that of the force due to the heave component of the motion. If the resultant control force is applied at the same point as the heave force no pitch will occur. This position is termed the 'neutral point' of the submarine. (Figure 8.7) A control force acting at submarine without any change in its pitch angle, thus enabling it to glide up or down while remaining horizontal. To obtain the largest possible ratio of pitch angle to control surface deflection, the M_δ/Z_δ term should be as negative as possible, i.e. the control surfaces should be as far aft as can be contrived (which is just like the situation for the rudder for motion in the horizontal plane).

It should be noticed that, apart from the relative location of the control force to the Neutral Point, the pitch angle depends on the ratio of control force to the pitch moment, i.e. Z_δ/M_θ.

Using hydroplanes as control surfaces the control force $Z_\delta \delta c$ will be of the form $C_l \frac{1}{2}\rho A U^2 \cdot \delta c$, i.e. Z_δ is a hydrodynamic force dependent on U^2. However, M_θ is the hydrostatic restoring moment which does not vary with speed. Hence the pitch angle caused by a plane deflection will vary with speed. On the other hand the heave velocity w depends only on the ratio of two hydrodynamic forces Z_δ/Z_α and hence for small motions it is independent of speed.

When the speed is relatively high, one would expect the submarine to pitch downwards due to an upward control force at the stern and, though this upward force would cause an upward velocity, the net result would be that the boat would incline downwards and dive down with only a slight upward drift angle from that direction; thus after planes only can be used to steer the submarine in depth at such high speeds. As speed is reduced, however, the pitch effectiveness term reduces in magnitude and at sufficiently low speeds the angle of pitch due to after plane operation can be less than the drift angle caused by the plane force. At such low speeds, when an upward force is applied by the after planes, the boat will pitch down but the dominant effect will be a bodily upwards movement. At some intermediate speeds there will be a circumstance in which an after plane force will cause a pitch angle which exactly matches the drift angle and then the submarine will be unable to change depth even though pitched (Figure 8.8). This is known as the 'critical speed', at which after planes are ineffective for changing

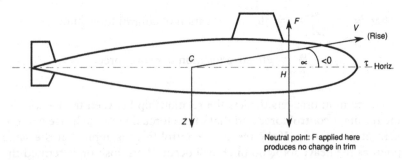

Fig. 8.7   Neutral point

depth, and then forward planes become necessary. The critical speed can be evaluated from the following formula deduced from the foregoing equations:

$$-\alpha = \theta$$

$$\therefore \; W\overline{BG} = Z_\alpha \, (x_\alpha - x_c)$$

If now Z_α is written in non-dimensional form $Z_\alpha' \times \dfrac{1}{2}\rho L^2 U^2$ we obtain

$$U_c^2 = \frac{W\overline{BG}}{\dfrac{1}{2}\rho L^2 Z_\alpha'} \, (x_\alpha - x_c)$$

The change of after plane control effectiveness at the critical speed is sometimes known as the 'Chinese effect'. The critical speed is typically around two knots or so and its significance lies in the fact that as it is approached with reducing submarine speed, the effectiveness of the after planes in controlling depth is progressively reduced.

It should be noted that with after planes the direction of the control force is in the opposite sense to the resultant pitch angle and the heave velocity component is in the opposite direction to that in which the pitch angle drives the submarine. With forward planes on the other hand, because they are forward of the neutral point, the control force is in the same sense as the pitch angle it causes, whilst the heave velocity component is in the same direction to that in which the pitch angle drives the boat. However, because the neutral point is forward, it is not possible to get the forward planes sufficiently ahead of the neutral point for good control effectiveness. Nevertheless, owing to the Chinese effect, forward planes are essential to good slow speed control at depth – and necessary for control at periscope depth, as they greatly assist rapid depth changing when a submarine dives from or close to the water surface.

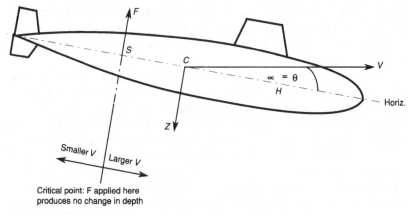

Fig. 8.8   Critical point

STEERING AND DEPTH CONTROL SYSTEMS

8.9 The control system in older submarines was essentially manual. There were three manned positions for the purpose in the Control Room: one was for the helmsman or coxswain with the steering wheel, and the other two were for the forward planesman and the after planesman. The information available to these control positions was provided by a compass, a pressure gauge for depth and a pitch angle (or trim) indicator, often in the form of a spirit level. The helmsman controlled or changed course, the forward planesman maintained or changed depth and the after planesman kept level trim or changed pitch angle. As indicated by the equations of motion previously developed, depth keeping or changing could be effected by operation of the forward planes at slow speed with relatively little effect on pitch and so that action did not interfere with the after planesman's task and similarly the after planesman's actions had little effect on depth, i.e. there was hardly any coupling in the activities relating to motion in the vertical plane. The way in which the planesmen operated their respective planes was by what is termed 'rate control', i.e. they determined the rate at which their control surfaces moved over and so each needed to have an indicator showing plane angle even though his ultimate function was to achieve ordered depth or pitch.

In modern submarines the manual operating mode has been reduced to a one man control console in which a single helmsman has a joy-stick and performs all three functions, an arrangement similar to that in aircraft. The joy-stick operates by what is termed 'position control', in which the plane and rudder angles match those to which the joy-stick has been moved. In combining course and depth control, the helmsman no longer has independent control of pitch and rate of change of depth and operates both forward and after planes together. There is usually provision for adjusting the ratio between the linked movements of forward and after planes and at higher speeds the forward planes can be disengaged (as the after planes are capable of depth control on their own) whereas at slower speeds higher ratio gearing of the forward planes is adopted. This procedure shifts the effective centre of control force to suit speed.

Whatever the way in which depth control is effected, it is possible for a submarine to take up a pitch angle while maintaining constant depth as a means of compensating for hydrostatic out-of-balance in either a force or moment sense (Figure 8.9). It is the function of the designated trimming officer to adjust the state of trim of the submarine using the trimming and compensating system. Although nuclear submarines as a matter of routine sustain sufficiently high speeds for quite substantial departures from trim to be hardly noticeable, it does necessitate running with plane angles set so as to offset the out-of-balance hydrostatic forces and moments hydrodynamically, and the effect is an undesirable increase in drag and noise. In battery driven submarines those consequences are intolerable and better practice of keeping in trim is essential – and customary.

8.10   The more pronounced coupling between pitch and depth control activities which occurs at higher speeds and the rather primitive instrumentation, made manual one-man control distinctly difficult. Efforts were made to ease the depth keeping task by introducing a form of display which combined actual depth and ordered depth so as to produce an error signal which the planesman aimed to reduce to zero, but these were tantamount to making him a simple servo in the control loop. The step that followed inevitably was the introduction of an automatic pilot system, and these are now in widespread use for all types of submarine.

Although auto-control systems are likely to become more sophisticated as control engineering advances, the majority of submarines currently in service operate with an auto-system based on the proportional integral, differential approach known as the 'PID system'. The simplest version involves a control surface deflection signal generated from a proportion of the error between ordered and actual position of the submarine, and an error rate term derived from the rate of change of that position – the latter is used to ensure that a stable and reasonably damped control is exercised. Thus for course control, the rudder angle signal is generated from the difference between ordered heading and actual heading plus a yaw rate term. For depth control, the control angle signal of the (usually linked) hydroplanes is normally generated from the difference between ordered depth and actual depth plus a rate of change of depth term. With some autopilots, the simplifying assumption is made that the submarine will travel along its axis and the depth rate term is

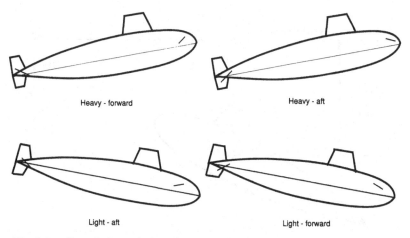

Heavy - forward   Heavy - aft

Light - aft   Light - forward

Fig. 8.9   Out of trim – pitch and plane settings

replaced by a pitch angle term. The effect of combining this pitch angle term with the depth error term can be regarded as equivalent to generating an imaginary depth error at some point ahead of the submarine which the control system operates to reduce to zero (Figure 8.10); the outcome, however, is not entirely satisfactory since if a change of depth is ordered, the approach to the new depth will be asymptotic in character and hence unduly slow. Although that response can be avoided by ordering a new depth rather larger than required (if increasing depth) it is clearly undesirable to have to fudge the process.

Another adverse effect can arise with an autopilot which uses a signal combining depth error and pitch angle, if the submarine is hydrostatically out-of-trim. If, for instance, the submarine is heavy and so adopts an upwards pitch angle to support the excess weight then when, say, an increase in depth is ordered, the submarine will move to an equilibrium depth condition below the ordered depth yet still pointing upwards so that the error as seen by the controller is zero. In this circumstance, the solution could be to order a new depth less than required so that the submarine does move to the required depth, but again that is undesirable. It is possible to design an autopilot to cope with a submarine being out-of-trim by incorporating an integral term in the depth error signal to be reduced to zero, but an integral term tends to introduce instability into control.

8.11 Another condition which places difficult demands on the design of an autopilot is control at periscope depth. As we have already observed, and return to again later, use of forward planes is essential to operation at periscope depth at which the submarine speed is unlikely to be much above 10 knots and could be much lower, but it remains a diffi-

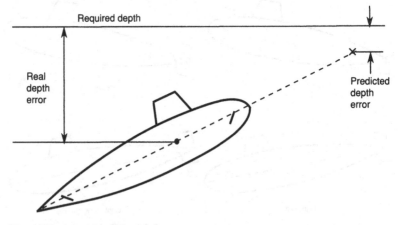

Fig. 8.10 Auto-depth keeping

cult task to achieve a reasonably clear view through the selected periscope when there are surface waves. Even if the surface of the water is relatively calm, the approach of the submarine towards the surface causes a suction force which the forward and after planes have to counter since the boat has to be kept level. With an autopilot the suction force really demands the inclusion of an integral term in the control algorithm.

With surface waves there is an additional effect at periscope depth, namely cyclic vertical forces and moments of a hydrostatic nature due to the variation in pressure on the hull of the submarine as the waves pass overhead. In a rough sea, waves of many frequencies occur but the submarine will only respond to the forces and moments due to the lower frequency waves, i.e. those of longer wave length. However, the pressure variations caused by all the wave components will be picked up by the depth sensor, which is usually a pressure gauge, and fed as signals to the autopilot – which, unless appropriately designed, will order the planes to respond to the apparent depth errors it seems to be detecting. The outcome would be unnecessary cycling by the planes because, rather as observed above, the submarine will only respond to the lower frequency disturbances. The solution adopted in most autopilots is to filter the depth sensor signal in such a way that only the low frequency components of the pressure variation being sensed are passed on to the plane actuators. Large groups of waves passing overhead also generate intermittent suction forces on the hull which are difficult to predict and counter with the control surfaces.

It will be appreciated from this account of operation at periscope depth that it makes quite severe demands on the autopilot. Interestingly, in older submarines with three-man manual control as previously described, maintenance of depth while using periscopes – given experienced helmsmen and planesmen – was quite good (because the men learned to disregard apparent fluctuations in depth which they could do nothing about) until the seas became too rough.

CONTRIBUTIONS OF HULL FORM AND APPENDAGES TO CONTROL DYNAMICS

The main hull

8.12   Although it is the control surfaces that apply the disturbing forces which initiate the change in path sought for by their operation, the major contribution comes from the hydrodynamic forces generated by the main hull itself. Thus the hull shape can significantly influence the submarine's control characteristics as well as its dynamic stability.

If the main hull were entirely axisymmetric, then in straight line motion in the direction of its axis no forces or moments would be generated by the hull. However, it is unusual for the main hull to be axisymmetric to that extent: though port and starboard symmetry is normal, few submarines are symmetrical about a horizontal plane through the axis (Figure 8.11). The most common feature is the addition of a casing above the main hull to provide stowage for equipments external to the pressure hull. If the casing is streamlined into the main hull, there is a tendency towards a cambered form such that, with the submarine moving on a straight path in the direction of its axis, a small lift force is generated upwards and a bow-up moment. The bow of the hull is rarely symmetrical top to bottom because of location forward of the main sonar, the torpedo tubes and the arrangement of the forward planes. Though a smooth faired shape is customary, the lack of symmetry results in a shift of the stagnation point at the bow which again gives the effect of camber to the hull shape (Figure 8.12).

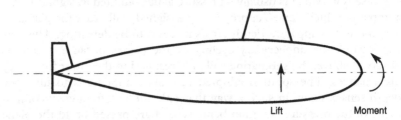

Lift   Moment

Fig. 8.11 Asymmetric hull form

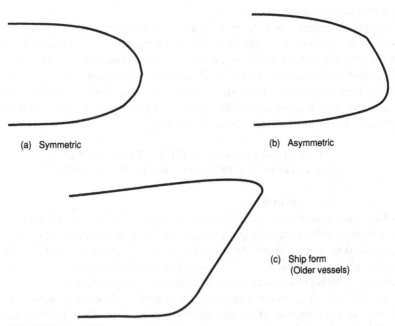

(a) Symmetric   (b) Asymmetric

(c) Ship form
     (Older vessels)

Fig. 8.12 Bow shapes

An extreme example of a very asymmetric bow is that of the old fleet submarines in which the provision of a ship-shape bow for surface running resulted in a sharp stem with flare. This flared and sharpened bow causes an upwards lift force at the bow when the submarine is in straight line motion with the keel level. Experience has shown that even minor variations in the way the bow of the submarine is configured as built can result in appreciable differences in the settings of the control surfaces for maintaining a straight and level path. Some hulls are fitted with a box keel which causes considerable appendage drag well below the axis and so leads to a downward pitching moment on the hull.

Most modern submarines have a symmetrical stern shape with the shaft for the single propulsor on the axis. However, on earlier submarines it was common practice to keep the top line of the pressure hull horizontal and 'cone up' the sections towards the stern, thereby raising the tail profile above the axis of the main hull. This resulted in an opposite effect to that of the casing, i.e. effectively leading to a downward camber generating a downward force.

The length to diameter ratio of the main hull can affect the turning and manoeuvring qualities of the submarine. As we have already commented, a streamlined hull form is dynamically unstable in both horizontal and vertical planes unless stabiliser fins and/or control surfaces are provided; the shorter and fuller the form is, the greater the degree of instability and the larger the fins and/or control surfaces required to impart dynamic stability. The short, fat submarine can in consequence be very manoeuvrable in both horizontal and vertical planes. For a short, fat submarine in a tight turn the angle of attack on the rudders may be negative, i.e. the rudders are actually holding the submarine back from turning as tightly as it would otherwise do. For longer, slender hulls, the tendency is for the unappended hull to be more stable dynamically and so require less fin and/or control surface area aft to achieve stability, but consequently the submarine will be less manoeuvrable – or require larger control surfaces to provide good manoeuvrability.

The sharp ship-shape bow of older submarines caused forces to be generated close to the bow when turning, due to the angle of attack at that point. The effect not only shifted the centre of action of yaw and sway forces further forward, tending towards greater instability, but also resulted in a shed vortex from the lower part of the hull which could impinge on and interact with after control surfaces, adversely altering their characteristics.

Bridge fin

8.13 In all hydrodynamic aspects of submarine design the existence of the bridge fin is totally undesirable. Nevertheless, virtually all designs require some form of faired large appendage on top of the hull to enclose the conning tower, the bridge conning position and the various periscopes and masts to be used by the submarine when at periscope depth. Despite efforts to minimise the size of this large appendage, it remains as a major excrescence causing considerable drag high above the axis so that in the vertical plane it causes an appreciable bow-up pitching moment. As we shall see, it causes many other adverse effects of which this asymmetric moment is not the most significant, though it is sufficient to override the other hydrodynamic effects on the main hull and so determine the settings required on the forward and after hydroplanes to allow the submarine to maintain a straight and level path.

When the submerged submarine turns in the horizontal plane, the action of the bridge fin is more complex. Efforts to reduce its drag result in a streamline shape, rather like the aerofoil section of a wing of low aspect ratio, which will generate lift when at an angle of attack to the flow. Those circumstances arise in a turn when the hull is in general drift motion and because the sideways lift force on the bridge fin acts well above the axis of the submarine a heeling moment results. The drift angle on a submarine in a tight turn can build up quite quickly and the dynamics of the corresponding heeling moment from the bridge fin can cause a 'snap roll', which although it is inward on the turn can be far larger than is appropriate for a banked turn effect (Figure 8.13). Even if not dangerous it can be decidedly disadvantageous to crew efficiency. There is consequently a need for an almost impossible hydrodynamic characteristic for the bridge fin, namely, to have both low drag and low lift. To a limited extent that can be achieved by having a fairly full section with a very blunt trailing edge, which gives the effect of stalling the section at low angles of attack, albeit with a penalty of increased drag on the turn.

Another way in which to reduce snap roll depends on the geometry of the turning path of the submarine. In the initial part of the turn when the rudders are put over, the first action is for the whole hull to go into a drift motion from whence the forces generated on the hull begin to cause the submarine to turn. Once the turn has been fully developed, the motion of the submarine is such that there is a variation in the angle of attack all along its length. With the attitude normally taken up by the hull, there is what is known as the 'turning point' about a quarter of the length back from the bow at which there is no angle of attack on the hull. Forward of this point the angle of attack is small and negative and aft the angle of attack is positive and becomes progressively larger further aft. To reduce the angle of attack on the bridge fin it is therefore desirable to locate it somewhere near the turning point of the hull, though that is likely to be further forward than is compatible with the internal arrangement of the submarine.

The lift force generated by the bridge fin also contributes to the sway and yaw forces and moments on the hull. The longitudinal location of the bridge fin therefore influences the points of action of the sway and yaw forces and so affects dynamic stability in the horizontal plane, usually in such a way that if it were located well forward at or near to the turning point dynamic stability would be much reduced – and the further aft its location, the less would be the reduction in dynamic stability.

An even more complex effect of the bridge fin is associated with its low aspect ratio. When the hull is at an angle of drift in a turn, the bridge fin in generating its lift force will be shedding a strong vortex at its top (Figure 8.14). The trailing tip vortex so formed not only modifies the angle of attack on the bridge fin but also induces a cross-flow on the main hull, astern of the bridge fin, modifying the turning forces on the hull. Quite apart from that effect, the hull itself being at an angle of attack will shed vortices from the top and bottom. The effect of the bridge fin vortex and its corresponding image in the hull is to modify the top and bottom vortices shed by the main hull. The outcome of the interactions of these vortex flows is the generation of a force and moment on the hull in the vertical plane, which would cause it to pitch and change depth if no control action were taken. The most common form of this complex interaction is for the submarine to pitch bow up and sink in depth, the so-called 'stern dipping' tendency. Not all submarines behave in that way, however, and it has been known for the opposite effect to happen, i.e. a bow-down attitude and a rise in depth, or even for boats to have those characteristics differently for port and starboard turns. No simple explanation can be offered, except that the different actions could

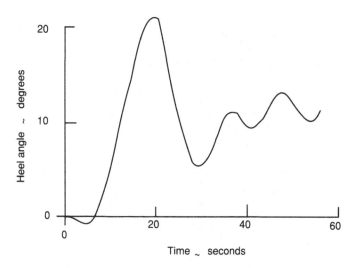

Fig. 8.13 Heel angle on turn

be due to the configuration of the casing on the hull as they have been found to be much more pronounced in boats with a substantial casing than those which are more nearly axisymmetric. Relatively slight geometric differences between port and starboard in submarines as built – which do happen – could be sufficient to account for the difference in behaviour between port and starboard.

At the base of the bridge fin where it intersects the hull, the possibility arises of other vortex generation. At the forward end there can be a local stagnation point from which separation of flow can occur. This takes the form of a pair of opposite handed vortex filaments passing down either side of the fin and over the stern casing. This is often called a 'horseshoe' vortex. It is not thought to have any significant effects on control but does increase drag and further complicate the fluid flow over the stern and into the propulsor.

Forward hydroplanes

8.19   There is even today much debate and wide differences of opinion on both the need for and location of fore hydroplanes. As we saw in the section on submarine manoeuvring characteristics, the forward planes are useful at relatively low speeds and at periscope depth. At high speeds, pitch angle and depth can be controlled adequately using only the after planes. Fitting fore planes does provide the means of controlling pitch angle and depth independently, making it possible for the submarine to remain level while slowly gliding up or down in depth. That manoeuvre can be achieved by setting the fore and after planes so that the net control force effectively acts at the neutral point, causing it to rise or sink

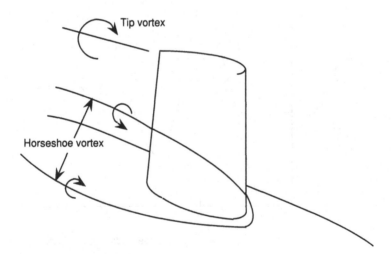

Fig. 8.14 Bridge fin vortices

without pitching moment. This capability has led some designers to see virtue in fitting the fore planes at the neutral point so that the submarine can be made to rise and sink by action of those planes alone.

A further development of that approach has been to mount the fore planes on the bridge fin in those submarines in which, for internal arrangement reasons, the bridge fin is located well forward and thus near to the neutral point (Figure 8.15). Because of the relatively narrow beam of the bridge fin it is possible for the fore planes to have a high aspect ratio without extending beyond the maximum beam of the submarine. That location also serves to remove the fore plane operating gear from the bow, so reducing the self noise penalty to sonar. It also avoids the need for retraction to avoid damage when coming alongside.

Against those advantages of the bridge fin location for fore planes, however, there are some disadvantages. The planes are high up and so their drag adds to the pitching moment on the submarine when level. When the submarine approaches the surface, the planes are in greater wave action than if they were on the casing or even lower on the external hull forward, adding to the difficulty of keeping periscope depth. When the submarine is on the surface, the planes are high above the water and are of no help in aiding it to dive quickly from the surface.

When located as far forward as possible, the fore planes not only generate a rising or diving force but also a pitching moment in the same sense. This feature serves to provide a counter to jammed after planes in an emergency situation, though forward planes wherever located are not sufficient to cope on their own with other than a relatively small jam angle. The arrangement with fore planes located well forward also has disadvantages. From a hydrodynamic point of view it is desirable to mount the planes at mid height, level with the axis of the hull, where the flow over the planes is symmetric enabling them to be equally effective

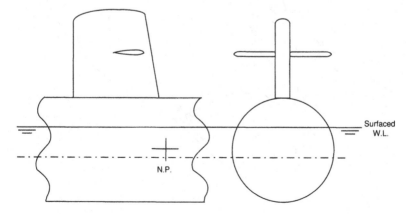

Fig. 8.15   Bridge fin planes

for rise and dive manoeuvres; planes in that location are termed 'drowned' planes and are also good for diving from the surface. To give them reasonable aspect ratio there they must extend well beyond the maximum beam of the main hull and provision for retraction is inescapable. Retractable planes have the virtue that they can be withdrawn at higher speeds, reducing drag and flow noise near the bow sonar, though if retracted they are not available for emergency recovery purposes, and provision of the ability to retract does add considerably to the complication of the operating gear and also demands room in the external hull forward where space is at a premium.

There is a hydrodynamic effect on the after planes associated with drowned forward planes. When the foreplanes are operated, they shed a tip vortex which streams down the hull and then over the after planes (Figure 8.16).

Consequently, action of the forward planes both generates a lift locally and causes a small negative lift effect on the after planes, which can result in the effective point of action of drowned forward planes being less easily determined for control purposes than if they were located elsewhere, where the trailing vortex did not interact with the rest of the hull and the after planes. For these several reasons, fore planes are not invariably sited at axis level. One possibility would be to mount them even lower

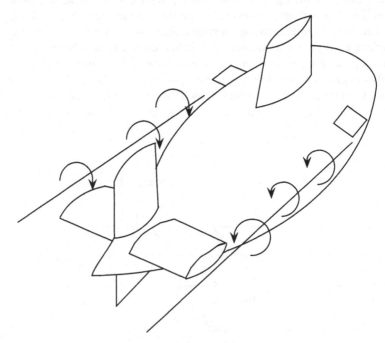

Fig. 8.16 Fore plane interaction on after planes

down on the hull towards the keel where they can project outwards without extending beyond the maximum beam and so need not be retractable; however, the angle of intersection of the plane axis with the hull is acute, making it difficult to derive a suitable geometry for this alternative. More commonly, the forward planes are mounted on the casing forward, which enables the operating and retracting gear to be housed within the casing where space is not at such a high premium. Casing mounted planes are necessarily located further aft than drowned planes and in addition, as with underslung planes, they intersect the hull at an angle which causes the local flow to be very asymmetric. It is usually necessary to contrive some fairing to the upper hull area to provide a relatively flat surface on which the inner edge of the planes can rotate – otherwise there is a gap where the inner edge has to be cut away to clear hull and casing. Usually the local flow in the region of casing mounted planes is upward, which means that when they are set horizontal they are actually at an angle of attack giving a rise force. To cater for this effect, the planes have to be set at a small angle known as the 'zero lift angle' in which they are aligned to the local flow.

After hydroplanes

8.15   The after hydroplanes are the horizontal control surfaces situated as far to the stern of the submarine as is practicable which, in the case of a single propulsor with its shaft on the axis, usually means that the planes are at the level of the axis just ahead of the propulsor on the tail cone of the hull. The main function of the planes, as we have seen, is to control the pitch of the submarine – though, as indicated by the equations of motion, at higher speeds their actuation to control pitch also determines how the boat rises or dives, so that it can be steered vertically using after planes alone.

Because it is, in general, desirable to have a degree of dynamic stability to motion in the vertical plane, the submarine needs to have quite large horizontal surfaces (whether planes or stabiliser fins) at the stern. As would be expected, the aim in design is to keep those after surfaces within the maximum beam of the submarine, for coming alongside reasons, but – depending on the cone angle of the hull at the stern – this may not allow a reasonable aspect ratio to be achieved and extension beyond maximum beam will then be unavoidable. In those circumstances, since it is not feasible within the already crowded external hull space aft to accommodate the means of retraction, it is likely that some part of the after surfaces will be made structurally rigid so as to be able to withstand impact when coming alongside.

The fairly large area desirable for the horizontal surfaces aft is associated with dynamic stability but is not necessary for depth changing and keeping. Thus a common design compromise adopted is to provide flaps at the after edge of fixed stabiliser fins. The proportion of flap area to total horizontal surface area can then be varied to suit the designer's intentions as to the appropriate balance between stability and manoeuvrability in the vertical plane. Control of depth at high speeds calls for only a relatively small control force at the stern, but at slow speeds depth control requires a relatively large force to overcome the hydrostatic restoring moment opposing change of pitch angle (Figure 8.17(a)).

A compromise is usually reached in which the after plane area is determined to suit low speed control and to limit the angle to which the planes

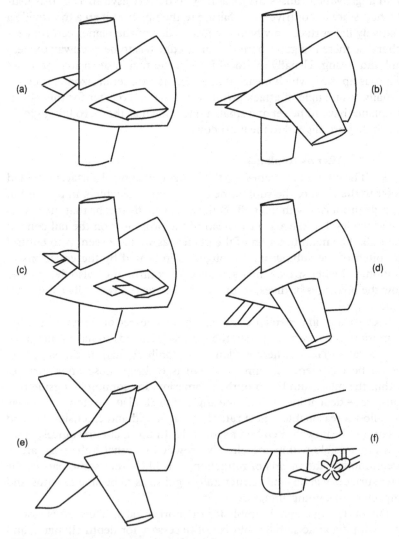

Fig. 8.17 Alternative control configurations

are allowed to operate at high speeds. That solution calls for an accurate operating mechanism with very little backlash. An alternative approach, not in common use, is to split the flaps into two sections either side of the axis and arrange the operating gear so that at high speeds only one pair of split flaps are used while both pairs are used at slow speeds (Figure 8.17 (c)).

In older conventional submarines, the functions of stabilisation and control of depth were independent, with the after planes and fixed stabilising fins being separate. With the ship-shaped stern and twin propeller configuration of those boats, the arrangement led to the stabilisers being mounted high on the hull and the after planes mounted on the rudder skeg (which acted as a vertical stabiliser fin). So located, each of the after planes was in the propeller race, which gave them good effectiveness, particularly at slow speeds (Figure 8.17(f)).

One of the design problems encountered in modern submarines with a single propulsor on the axis is the difficulty of achieving a continuous stock connecting the after planes together (the same difficulty occurs with rudders located in the vertical plane through the axis). The solution is to split the stock into two separate lengths and to contrive a yoke which goes around the axial shaft and connects the two lengths together, but, in consequence, the external hull aft is crowded with operating gear, a fact which discourages adding to the complication either by split planes, as described above, or by making the port and starboard planes separate and capable of being angled differentially to counteract snap roll on the turn due to the bridge fin.

In the single propulsor arrangement, with the after planes mounted on the stern ahead of the propulsor, little advantage is gained from the race. Moreover, the planes and stabiliser fins (and also the rudders) create a wake shadow into the propulsor which causes variations in angle of attack of the rotor blades, an adverse effect for both noise and vibration reasons. Deflection of the control surfaces aft causes an increase in wake disturbance and to minimise this adverse effect the designer will aim to keep the trailing edge of the planes and rudders a reasonable distance ahead of the propulsor.

At high speeds the after planes dominate manoeuvrability and control in depth and it is important that their operating gear should be very reliable, so that the risk of a failure is remote and will not cause an uncontrollable dive (inadvertent rise is also undesirable, but not potentially so catastrophic). A possible cause for concern in this regard is that the planes might become disconnected from the actuating system, and to counteract that occurrence the designer should try to ensure that the planes would then trail at only a small angle of incidence by keeping their area balance to a minimum, i.e. arranging the tilt axis as close to the leading edge as possible. A contingency considered to represent a greater

risk is an after plane jam and we describe in later sections how that hazard can be catered for by emergency recovery action or could be avoided
by redesign of the arrangement of the control surfaces.

After rudders

8.16   In submarines with a single propulsor the normal arrangement is
to have a pair of vertical rudders one above and one below the tail cone
more or less in line with the after planes and fins. The total area of the
rudders is determined by dynamic stability considerations in relation to
course keeping, but it is less necessary to retain stability on the turn, and
so to achieve fast turning the whole of the vertical surfaces are made
moveable. It is difficult to contrive a very effective lower rudder due to
the limited space between the tail cone and the line of keel. In the double
rudder configuration it is therefore quite usual for the upper rudder to be
higher, i.e. with a larger span and with a larger area than the lower rudder. When the submarine is submerged that arrangement is quite satisfactory and has an advantage in that the roll moment caused by the
asymmetrical rudders counteracts the roll moment due to the bridge fin.
However, when on the surface the large upper rudder is out of the water
and so is ineffective for steering the submarine, which can lead to rather
poor handling qualities. In older conventional submarines with a ship-
shape stern it was the practice to have a single rudder between the twin
propellers, which was of the flap type behind a fixed skeg on which the
after planes were mounted.

The combination of propulsor shaft, hydroplane and rudder stocks in
the submarine with a single axial propulsor leads, as we just observed, to
severe congestion in the external hull aft. Because the rudders are all-
moveable surfaces, the centre line of their stock can be further forward
than that of the flapped after planes and so can be separated from it longitudinally, but it is still necessary, if the upper and lower rudders are to
be operated together, to provide a yoke around the main shaft. Another
consequence of all this massive equipment so far aft in the submarine is
that it can contribute to difficulty in achieving hydrostatic longitudinal
balance.

Alternative stern configurations

8.17   The configuration at the stern with vertical rudders and horizontal fins and planes is known as the cruciform arrangement (Figure
8.17(a)). Because of the difficulties of achieving after lifting surfaces of
suitable aspect ratios with that arrangement, alternative configurations
have been developed and even fitted in some modern submarines. The
most common of these is known as the 'X' stern (Figure 8.17(e)). In that
arrangement, two pairs of control surfaces are used, arranged at 45° to
the horizontal and vertical planes through the axis of the hull. With this

angling it is possible to get a much larger span for each of the surfaces while not exceeding the limits of the box defined by the maximum beam and draft of the hull. Usually the pairs of surfaces are cross-connected through the axis with a common stock so that each pair moves in unison, though differently from the other pair. To achieve the control forces for turning or diving a combination of movements of the pairs of control surfaces is required. Thus to turn the submarine both stocks are rotated in the same sense so that the forces generated by the surfaces have components which add together in the horizontal direction but cancel each other out in the vertical direction. To dive the submarine the stocks are rotated in the opposite sense so that the horizontal components cancel out and the vertical components add together. Control with the X stern configuration is only really suitable for use with one-man control systems in which the servo-mechanisms are so designed that the required differential rotations are contrived automatically. The helmsman only needs to act as if he is engaged in turning/course-keeping or diving/depth-keeping in the ordinary way.

Apart from the advantage of better aspect ratio for the control surfaces the X stern has other virtues. One is that there is a pair of fully effective lower control surfaces when the submarine is running on the surface so that steering is improved. Another is that, should the stock for one pair of control surfaces jam, it is possible to change the angle of the other pair in such a way as to cancel any diving or rising action and cause the submarine to turn, so avoiding a potentially dangerous change of depth.

There is, in our view, a disadvantage with the configuration. The forces, generated in both vertical and horizontal directions, are symmetric and there is little scope for independent adjustment of the dynamic stability and control characteristics of a submarine for turning/course-keeping and for diving/depth-keeping purposes. The former calls for relatively low stability/high turning forces and the latter relatively high stability/low diving forces. If the control surfaces all rotate to the same extent, then the amount of deflection which causes the submarine to turn tightly will – unless inhibited in some way – cause an undesirably violent response in the vertical plane when they are operated in that mode. It follows that it is necessary to provide the means of limiting the action of the control surfaces in the X configuration so that they cannot be used to their full extent for motion in the vertical plane.

Another problem with the arrangement is that if it were entirely symmetrical the axes of both stocks would pass through that part of the axis of the submarine occupied by the propulsor shaft, resulting in congestion and interference. To avoid that consequence, one solution is to displace one pair of surfaces longitudinally relative to the other pair sufficiently to give clearance between the yokes required for the shaft to pass through. Another solution is to provide independent means of operation for all four control surfaces, but that arrangement adds further to the complexity of the servo mechanisms required for control. It does, however, allow differential movement of the control surfaces to counteract snap roll on turning.

8.18   The inability to tailor stability and control in horizontal and vertical planes with the X stern have led to consideration of alternative configurations. One is the inverted Y stern, (Figure 8.17(d) in which there is a single vertical rudder above the hull and two control surfaces each at 45° below the hull. This arrangement allows the advantage of high aspect ratio to be retained while still giving two control surfaces fully immersed acting as rudders when the submarine is running on the surface. With the inverted Y stern, as with the X stern, it is necessary to contrive the control system so that the movements of the control surfaces are so co-ordinated as to enable them to act just as rudders or just as after hydroplanes as required. The arrangement does lead to more complexity in the control system as there is no commonality of the stocks and it is necessary to provide three separate actuating mechanisms for the three independent surfaces. The arrangement does enable the designer to provide sufficient area effectively in the vertical plane to give low dynamic stability/high control forces in the horizontal plane while at the same time allowing high stability/low control forces in the vertical plane. The danger of overpowering in relation to diving motions is avoided because the vertical rudder provides the extra power for turning motions with hardly any impact on diving.

Yet other alternatives to the X or Y stern configurations have been considered which use variations on the geometry, for example, to have a vertical pair of rudders and a pair of planes which are not horizontal but angled down at an anhedral angle (Figure 8.17(b) – the aim being to retain most of the advantages of the inverted Y stern while retaining the through stock for normal operation of the rudders. As things stand at present the cruciform arrangement of control surfaces aft predominates and the hazards are coped with as described below in the section on emergency recovery.

Propulsor
8.19   We have described the issues which bear on the choice of propulsor arrangement and for now concentrate on its relevance to dynamic stability and control. Although the primary purpose of the propulsor is to provide axial thrust to enable the submarine to achieve the intended ahead speed, it should not be overlooked that it also has an important impact on the dynamic stability and control characteristics in both horizontal and vertical planes. The action of the propulsor is to induce an increase in longitudinal velocity of the water flowing through it, thereby

changing its momentum and so providing the thrust. However, when the submarine is in sideways motion (whether vertically or horizontally) the inflow to the propulsor is no longer axial and the action of the propulsor is modified so that it causes an angular change in momentum of the water, thus generating a sideways force. It thereby effectively acts as a stabilising fin at the stern of the submarine. A large diameter single propeller can in that way have a significant stabilising effect, but a smaller diameter propeller or pump jet will have less effect because it works in a wake field which has already been re-aligned due to the action of the hull.

EMERGENCY RECOVERY

The hazards

8.20   The impact on submarine design of assurance of safety of submerged operation is a large subject which embraces all of the considerations we treat as chapter headings in this book, but none is so profoundly involved as the set of issues we address in the present chapter. Apart from fire, the greatest hazards a submerged submarine faces are from flooding at depth and from uncontrollable diving. The former, unlike the latter which is obviously a dynamics phenomenon, might seem out of place in the present context as it is a hydrostatics phenomenon -, but, as we shall see, the road to recovery lies partly in invoking hydrodynamic forces in an effort to override threatening hydrostatic forces. The present section is, therefore, concerned not only with the nature of the threats to safety but also with what actions can be taken to aid emergency recovery.

Necessarily, emergency recovery is a matter both of what provision to make in design and what action to take in operation and the two aspects are inextricably interlocked. In what follows we concentrate attention on the design aspect because this book is about design, but that in no way diminishes the importance of the operational aspects. It is the limitation of time and space which necessitates our treating operational considerations just to the extent required to represent properly the design considerations.

The flooding incident

8.21   Since the first submersibles, avoidance of inadvertent flooding has been the pre-occupation of the submarine designer – and of the submariner – because of the characteristic small reserve of buoyancy on the surface and total lack of reserve submerged. Flooding can be caused by any one of a number of incidents: maloperation like leaving hatches improperly secured; failure of piping exposed to sea pressure; piercing of the pressure hull in collision or by enemy action; by poor design of pressure boundary interlocking arrangements intended to prevent accidental ingress of water; or by poor quality control of sea pressure-carrying equipment during construction or refit. While any such incident would be an embarrassment if it occurred with the submarine on the surface it is potentially catastrophic when submerged, and the deeper the submarine the greater the risk. Even a small hole of a few centimetres diameter at full diving depth could lead to a rapid rate of flooding and would be accompanied by blinding spray capable of shorting out any electrical equipment in its path. The only course of action would be to get the submarine back towards the surface as quickly as possible to enable the crew to escape – with most modern submarines the chances of saving the boat from subsequently sinking are negligible.

As discussed in an earlier chapter, an important step in emergency recovery from a flooding incident wherever along the length of the boat it occurred would be to blow all main ballast tanks, i.e. to invoke as much buoyancy as could be mustered by expelling water from those tanks. For most submarines the way to do that would be to apply a full HP air blow to the MBTs. At or approaching full diving depth even all the stored air would not immediately empty the tanks but would form pockets of air which could expand once the submarine started to rise towards the surface. That is when hydrodynamics could play an important role in aiding emergency recovery because, if the submarine were able to increase its ahead speed, the forward and after hydroplanes could be used both to help drive it upwards and to help counteract the increasing negative buoyancy and its moment. The corollary is that a submarine cannot afford to proceed too slowly when it is operating deep and it is vitally important to aim in design that even during a severe flooding incident propulsive power should not be lost. Clearly, one of the main objectives should be to limit the extent of the internal piping subject to external sea pressure. Any such internal sea water systems should be fitted with power operated hull valves that can be rapidly shut to isolate an internal failure and limit the extent of flooding. The operational philosophy of such valves requires careful evaluation as the shutting off of cooling systems may result in an automatic safety shut down of power generation systems, denying the vessel the capability to drive itself out of trouble due to the flooding.

The depth excursion incident

8.22   Concern about the risk of a submarine taking an unintentional excursion in depth which will prove difficult to control is more recent than that about inadvertent flooding. It is more characteristic of the high speed submarine both because of the shorter time in which recovery action can become effective and because the hydrodynamic forces can swamp any hydrostatic forces (like blowing MBTs) which can be quickly invoked. To illustrate how short the time scale may be, a submarine travelling at a speed of 30 knots would transit through a depth band of, say, 300 metres in under a minute, if it had a pitch angle of 20°.

As discussed in the earlier section on control surfaces, the incident which is considered as most hazardous as regards inadvertent diving at speed is jammed after hydroplanes. Unlike accidental flooding, which has occurred on several occasions in submarine operations, there is little evidence of after planes jamming in service, though there is some of the lesser incident in which they become disconnected from their operating gear. Thus the jammed after planes case should be regarded as a 'maximum credible accident'. The problem is to decide what angle of jam is credible. Current UK practice is to assume that the planes will jam at whatever angle they have been set to, for example, that having put the submarine into a dive with 10° of plane angle it is found that this angle cannot be taken off. A more severe accident assumption is that the planes will move over to maximum angle and jam irrespective of the manoeuvre being conducted. The first assumption permits the application of limitations on plane movements depending on depth and speed to avoid large excursions. The latter assumption gives no scope for the application of such operational guidance.

With an after plane jam at sufficiently high speeds, the forward planes – if they have not been retracted – would counter but not override the after control moment. The significance of the incident as a hazard to be taken account of in design is that it emphasises the importance of reducing ahead speed as much and as quickly as possible. Time is of the essence and whilst it would be unlikely that the propulsor could be put in reverse in the available time – which would have considerable braking effect – to reduce revolutions would help in slowing down. Another contribution to braking effect would be to put the rudders over, as speed reduction in turning is substantial, but the acceptability of that measure would depend on whether the submarine was known to exhibit the 'stern dipping' tendency or not.

It follows that for emergency recovery from jammed after planes there is a sequence of actions to be taken in whatever order proved to be practicable: reduce speed; put forward planes hard to rise; blow MBTs; put rudders hard over. But the main conclusion to be drawn from consideration of the incident is that the problem basically stems from high speed and that speed must be progressively restricted as the submarine dives deeper – because that leaves a progressively smaller margin between present depth and full diving depth. Further, on the basis that the after planes will not jam at an angle of tilt larger than that when the incident occurs, the allowable angle should be progressively reduced as speed is increased. The foregoing is related to the assumption of a cruciform stern arrangement, but even with other configurations some limitations on speed and control surface angles seem sensible.

Manoeuvring limitations:

8.23   As would be expected, the existence of hazards of the sort described above calls for the issuing to the submarine operators of advice on appropriate emergency recovery actions in association with rules on how the speed of the submarine should be limited according to its operating depth and the associated limits on permissible after plane operating angles. But, as we have seen, the limits are different according to the type of incident to be catered for: flooding at depth requires a minimum speed restriction, while jammed after planes requires a maximum speed/plane angle restriction as a function of depth.

The usual way in which to convey that information to submariners is by the use of manoeuvring limitation diagrams (Figure 8.18). The submarine commander is authorised to depart from these diagrams in special operational circumstances, but generally has to comply with them.

IMPACT ON DESIGN

8.24   As already observed, safety considerations loom large in design and pervade all aspects, and for now we will concentrate on those relevant to dynamics and control. There is, as so often happens in design, conflict between some of the pertinent considerations. Thus with the cruciform stern arrangements, the after planes, if jammed, bring risks of depth excursions at speed which call for a number of responses which fully extend recovery capability; so why have such large after planes? Yet

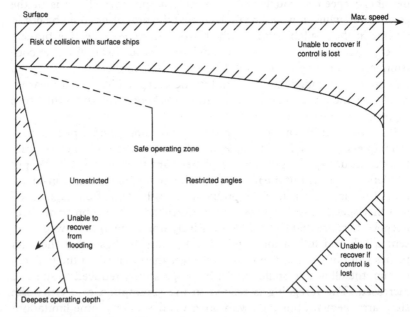

Fig. 8.18 Submarine safety envelope

for recovery from a flooding incident located aft in the submarine large after planes can make a very welcome contribution and, as we have seen, to change to an X stern configuration would be a mixed blessing.

So consideration of catering by other means for malfunction of the after planes in the cruciform arrangement is only to be expected, e.g. by providing an HP air cylinder to by-pass the hydraulic jacks in case they fail; or by having mechanical stops inserted to limit after plane angle above a certain speed (and so guarantee the magnitude of the maximum jam angle; or by having split after planes with only half normally in action at higher speeds; or a combination of some of these alternatives.

Another design aspect, as we have seen, is the choice of appropriate area of the horizontal stabiliser fins aft: too much would produce a sluggish response in depth changing and unwelcome stabilisation to a dive path; too little would lead to difficulty in depth keeping at speed, although stability (though not in a 'stick fixed' sense) could be incorporated in the autopilot controls.

Then there is the matter of the area, shape and location of the bridge fin and its adverse effect in causing snap roll – another safety consideration.

By and large, although dynamics and control are not major determinants in submarine sizing, they do significantly influence the provision for handling of the submarine and its inherent safety. The converse is also true, that the choice of hull and appendage configuration can have a significant influence on the natural dynamic behaviour of the vessel and lead to differing solutions of the control requirements.

9 SUBMARINE SYSTEMS

INTRODUCTION

9.1 Whilst all vehicles and ships have operating systems, the submarine requires special systems to enable it to operate in its environment between the surface and fully submerged below the sea. The systems are required both for operation in underwater space and for the crew to work efficiently totally divorced from the atmosphere. In concept formulation and initial sizing the systems do not present issues that need to be considered in detail, though they do require provision of space and weight within the hull. Nevertheless, since they are so important to the general operation of the vessel, we devote this chapter to describing particular aspects of submarine systems, how they are operated and the usual form taken by their design. A particularly important aspect of the systems is their integrity and reliability.

The primary systems in a submarine can be categorised under the following headings, though in some cases they overlap:

(a) **Hydraulic systems** – these are provided for power actuation of many valves, systems and controls.

(b) **High pressure air system** – this is required mainly to initially discharge water from the main ballast tanks in changing from submerged to surface condition, but it also has many other uses on board.

(c) **Water distribution systems** – these are required to control the trim of the boat and to eject unwanted water taken on board during various operations.

(d) **Ventilation and air-conditioning systems** – these are distinctive in a submarine because of the special needs of the enclosed atmosphere once submerged.

(e) **Electrical power distribution system** – nearly all the above systems at some stage require electrical power and it is therefore an all pervading system throughout the boat.

Electrical power affects virtually all the other systems, and the provision of energy storage and generation of electrical power are essential to the operation of the boat. However, some equipments are provided with a means of manual operation as a fall-back position in the event of loss of power.

HYDRAULIC SYSTEMS

9.2   There are so many requirements on board a submarine for mechanical actuation of equipment, often calling for considerable force, that it is common practice to provide a centralised hydraulics system distributing hydraulic power throughout the boat. The primary use of large amounts of hydraulic power is for the actuators of the control surfaces, the rudders and hydroplanes. Next is the requirement to raise and lower periscopes and masts in the bridge fin, and for a number of hull valves which require power as they operate against sea pressure. Some of the machinery systems also require hydraulic operating systems, though these may be independent. Some care has to be exercised in the use of hydraulic power in the engine room and hot spaces of the boat as a hydraulic leak (which, due to the system pressure, can result in a spray of the oil-based fluid) may cause a fire and so, in some vessels, a special non-inflammable fluid has been used in machinery spaces. Because of the presence of a very capable hydraulic system in a submarine it may be used for functions where it is not necessarily used in a surface ship; an example is for winches which have to operate underwater.

In earlier boats the hydraulic system was termed the 'Telemotor System' and this was a relatively low pressure system, 1500psi/100bar, which provided hydraulic fluid on demand. Essentially in the operation of that system the necessary valves were opened to direct the movement of the hydraulic fluid and a telemotor pump, normally electrically driven, started up to supply fluid to the selected ram or actuator in order to move the device. This process did, however, involve the frequent starting and stopping of the pump, which in a submarine is undesirable because of the associated noise.

It is now more common to adopt a hydraulic system which provides essentially a constant pressure supply. In such a system the supply lines throughout the boat are maintained at a high pressure, 3000psi/200bar, by means of pressure accumulators, which provide capacitance in the system. Pumps in the system top up the accumulators when they begin to exhaust their contents. By this means, the pumps are not always operating and the boat can continue to function for some considerable time on the stored energy within the accumulators. The general arrangement of such a centralised constant pressure system is to provide one or more supply pipes throughout the boat from which tappings to various equipments are taken, and the hydraulic oil is exhausted from the actuators into a return line which takes it back to the centralised system (Figure 9.1). Located in the boat, usually in the auxiliary machinery space, is a hydraulic pressurising plant consisting of a number of pumps and accumulators providing the pressure to the system. The used fluid is returned to a storage tank which feeds the suction of the pumps.

The lower pressure of the earlier systems necessitated very large rams and actuators in order to provide the forces and power levels required. However, they were relatively quiet systems as the velocities of oil involved were quite low. The higher constant pressure systems now used reduce the size of the mechanisms and actuators, making them more compact. Relatively large piping is still used to keep the velocity of the system low, giving relatively low flow noise and low piping losses. There are some arguments in favour of introducing even higher pressure systems, which have been used commercially, giving yet smaller sized components and piping, but the noise penalty with such systems has led submarine designers to avoid them.

9.3 In the old telemotor systems and in some parts of the more recent constant pressure systems, actuation is based on rate control. This involves the direct opening of a control valve to the selected actuator, the amount of opening determining the rate of the motion of the actuator. The problem with the technique is that it has to transmit information on the position of the actuator back to the operator who is opening or shutting the control valve. For some applications, such as raising a periscope, it is visually obvious what position the periscope is in at any stage and so the operator in the control room can stop and slow the mast as it reaches the required height. But with the actuation of control surfaces, by means of piston ram type actuators, these are sited at the ends of the vessel and are not directly visible to the operator. It is consequently necessary to provide feedback of ram position to the control room so the operator can know what position the hydroplanes or rudders are in. Problems arise with that method because, although it is possible to transmit a signal on the position of the ram, the actual position of the control surface

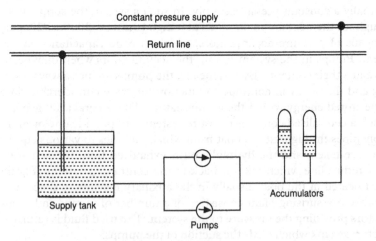

Fig. 9.1   Basic constant pressure power plant

depends on the linkage between the ram inboard and the shaft of the control surface outside the submarine. A situation has occurred where due to a fault those linkages have become disconnected and the plane operator believed he had positioned a control surface because of the information on the ram position when in fact the control surface was not responding. Another disadvantage is that rate control tends be noisy because the actuator is driven to near its required position and then shut off against the pressurised flow.

Modern systems employ what is known as position control. (Figure 9.2) With that technique the operator demands the required displacement of the actuator by means of a lever or dial and a servo system opens up a valve at the actuator to allow oil to flow until the motion of the actuator causes the valve to shut as it reaches its demand position. This is a relatively simple position error servo method. With the higher pressures in constant pressure systems the method can result in a high throttling noise when the valve slowly closes as the position error decreases. Nevertheless, position control has several advantages. It only requires a local take off of pressure supply, whereas rate control requires piping from the control position to the actuator, so increasing the number of pipes running through the boat. The operator has only to select the required outcome rather than having to judge the amount of oil to allow through the system. Position control can be effected by small electrical cables transmitting a signal to the local servo motor at the actuator. Due to the vital safety role of the control systems the servo distribution system is often duplicated in case of failure, while in some applications a fallback position of direct rate control is adopted, thus requiring additional piping. For the main control surfaces other fall-back positions are often adopted using either high pressure air as discussed later or electrical drive systems or, as a final fall-back position, a local manual pumping system.

It can be appreciated that in devising the systems for a submarine it is all too easy to finish up with a multiplicity of through pipe runs in the

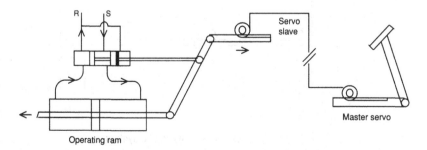

Operating ram

Fig. 9.2   Position control

boat. Even after making the effort to keep that tendency under control, it is still necessary to allow sufficient space for relatively large pipes with their couplings and valves. In particular, because many pipes have to pass through the watertight bulkheads, allowance must be made for the pipe penetrations.

9.4   The central hydraulic supply system consists of two or more electrically driven pumps delivering oil at pressure and connected to accumulators which are normally of the piston displacement type. The accumulator piston is backed on the other side by high pressure air supplied by local air bottles which need to be topped up from the air system from time to time. Switches are controlled by the displacement of the piston so that the pump is switched off when the accumulators are fully charged and only switched on when the accumulators are approaching the exhaustion point and pressure has dropped to the lower limit. The pumps take their supply of oil from local supply tanks to which the oil

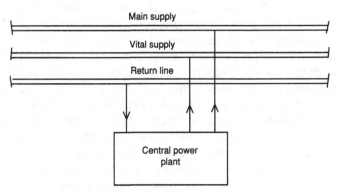

(a)   Central hydraulic power system

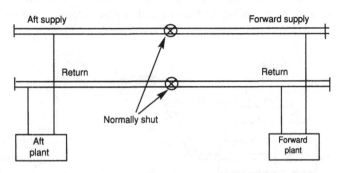

(b)   Fore and aft split hydraulic system

Fig. 9.3   Hydraulic systems

that has been used is returned. Thus the main piping through the boat consists of a supply line at constant high pressure and a return line at lower pressure leading to the supply tank. The number of pumps and accumulators adopted depends upon the demand of the system and also the failure philosophy of the design, in order at least to ensure that sufficient capacity is available in the event of breakdowns or maintenance.

As described before, the demand connections are tapped off the main supply line adjacent to the actuators of the particular mechanisms served, whether mast, valve or control surface. However, because of the importance of control surface actuation, it is common practice to provide for that purpose two supply lines, the main supply for general use by all actuators throughout the boat and a second supply dedicated solely for the use of the control system actuators. No other actuators are allowed to be connected to the latter. This arrangement helps to avoid the risk of loss of control due to malfunction occurring in another actuator. Thus it is quite usual in submarine design to provide duplicate supply piping systems running throughout the boat, each with dedicated accumulators and pumps (Figure 9.3 (a)). In normal operation these systems will be isolated from each other, though in failure conditions they may be cross-connected. This policy increases the space needed in the Auxiliary Machine Space (AMS) for the hydraulic system equipment and also increases the number of bulkhead penetrations and space used in duplicating the system throughout the boat. Together with direct rate control from the control room the hydraulic piping systems are in consequence quite a dominant feature in submarines.

9.5   Whilst it is more efficient to have centralised power systems, it does potentially compromise their integrity because local faults within the AMS may result in hydraulic power being lost throughout the boat. An alternative approach is to adopt a split system with one power plant towards the stern of the vessel and another one towards the fore end (Figure 9.3 (b)). Systems split fore and aft in that way enable shorter runs of piping with better integrity to the main control actuators and the provision of a single line throughout the boat for other hydraulic requirements. This single line through the boat can itself be split by an isolating valve near amidships, so that in the event of failure one or other parts of the system would remain intact. In the event of a failure somewhere along the main supply line it would be possible to isolate an element of piping associated with the failure and continue to operate forward and aft of the failure. It is conceivable that another alternative method would be to employ local hydraulic power packs, each directly associated with either one actuator or a group of actuators. For example, there could be a hydraulic pack associated with the after control systems located at the after end, and another hydraulic pack for the forward control surfaces located at the forward end, and these would either directly meet the local demand or act as a constant pressure system to meet other adjacent demands. Elsewhere, local power packs could provide power for other hydraulic power demands. This approach would increase the electrical distribution system providing power to the localised power packs, but would obviate the need for large diameter hydraulic piping at high pressure running throughout the boat.

Variations on these themes exist in submarine design. Some designs have a main hydraulic system with central power plant but a separate power plant located near the after end dedicated to the operation of after control surface actuators. The main supply provides an alternative back-up for local position or rate control.

9.6 The variety of alternative ways in which hydraulic systems can be designed has been emphasised because one of the main problems of those systems is the system integrity. There are numerous couplings and joints on the system which from time to time have to be broken for maintenance purposes and therefore it is necessary to provide high integrity joints, usually in the form of face or barrel O-ring design. Brazing of the joints is also adopted where there is no need to break pipes. In more recent designs welded joints have been adopted to increase the integrity, as brazing has posed problems of reliability mainly because, in the confines of a submarine, it is difficult to achieve uniform heating in the brazing procedure. O-ring joints are vulnerable to small defects or cuts which allow the leakage of oil, and once damaged they are liable to burst causing both a loss of hydraulic pressure and a spray of oil.

Another major consideration with hydraulic systems is the maintenance of system cleanliness. There is a danger of dirt and swarf being generated both in assembly during construction and due to wear of valves, which contaminates the oil and is liable to cause damage to O-rings and delicate valves in the system. The usual practice once the system is sealed is to flush it through to try to clear all the dirt out. However, because of the use of large diameter piping to reduce the velocity of flow, it is difficult to generate sufficient turbulence in the flow to satisfactorily flush the system. One means of overcoming that problem is to vibrate the piping while it is being flushed so that any pockets of dirt are shaken out of the system and washed through during the flushing.

9.7 A particular problem with oil systems in submarines is that there are actuators outside the pressure hull, e.g. to open and close shutters at the top of the bridge fin. The associated external system is prone to ingress of sea water into the oil, despite the positive differential pressure of the hydraulic system relative to sea pressure. Oil contaminated with sea water is a very difficult fluid to deal with as it tends to clog filters and is slow to separate, and so special steps have to be taken to protect hydraulic systems operating in sea water. In some designs the solution adopted is to provide an external hydraulic system separate from the normal main system. Even if this is not done, it is essential that separate return tanks are provided for external systems so that any sea water that leaks into the oil has an opportunity to separate out before it is carried over into the rest of the system.

The main actuators to the rudders and planes are usually of the piston ram type in which small valves govern the direction of the oil flow either side of the piston depending on the direction of motion required. The rams drive a sliding shaft directly through a pressure hull gland. External to the hull this shaft operates a crank lever to impart rotary motion to the control surface. The alternative of using vane type actuators for angular motion purposes has been considered, but there are reservations because it would necessitate provision of a primary hydraulic system external to the hull as the actuator would have to be mounted directly on the tilt shaft. This would pose problems both of sea water ingress and feedback to the position controller. The operation and raising of masts is also usually accomplished by a ram system, either with direct rams running the full length of the mast or with shorter rams operating through wires – the stroke of those rams generating sufficient motion in the wire to cause the full throw of the persicope or mast.

Power operated valves, such as the MBT vents and tank valves known as 'Kingstons' are of the mushroom type operated by a hydraulic ram, though some systems now adopt the full flow ball or cylinder valve requiring a rotating actuator.

HIGH PRESSURE AIR SYSTEMS

9.8   The primary purpose of high pressure air systems is to enable the vessel to surface, i.e. to displace water from the MBTs to change from the submerged condition to the surface condition. Though other ways are conceivable for displacing water from these tanks, the most common method is by means of high pressure air (HP Air).

As well as blowing MBTs, HP Air is used to discharge water from some internal tanks, in particular, the quick diving or negative tank fitted in many boats to help rapid diving from the surface and also D tanks fitted to compensate for the change in the buoyancy of the hull when going deep rapidly. In some designs HP Air is also used to transfer water between internal tanks in the trimming system and in older submarines to discharge water from the trim system direct to sea during the compensation operation.

Other uses for HP Air are for the operation of some pneumatic valves used in preference to hydraulic valves and for providing a salvage blow. The latter entails putting a pressure on a compartment inside the submarine to reduce flooding of that compartment, an operation which would only be carried out on or near the surface as it would be ineffective when deep. HP Air is also used in many submarines as an emergency means of operation of control surfaces by fitting an air ram as well as the normal hydraulic ram in the operating linkage.

Sizing the HP Air system

9.9   It is on the basis of fulfilling the primary purpose of the HP Air to blow water from the MBTs that the sizing of the system is established. In principle, water has to be discharged against sea pressure at the level of the flooding holes, which are at the bottom of the tanks. Assuming that the submarine has been brought to periscope depth, this pressure is approximately equivalent to three atmospheres; however, if the air pressure in the tanks was only of this level the initial discharge rate would be very low and the vessel would surface very slowly. As surfacing is a significant operation for the stability and safety of the vessel a larger differential pressure is called for to expel the water from the tanks rapidly. The HP Air has to be stored on board and occupy space within the outer envelope of the vessel, and it is advantageous for the air to be at as high a pressure as practicable so as to minimise its volume.

For a reasonably sized submarine, once it has achieved full surface buoyancy, the flooding holes are about 10 metres below the surface and the sea pressure at the flooding holes is about two atmospheres. Hence, for a complete blow of the tanks, a volume of air at atmospheric pressure is required equal to twice the volume of the tanks. However, if only this amount of air were carried a single surfacing would fully exhaust the system and the submarine would have to spend a considerable time on the surface, connected with the atmosphere by its ventilation mast, in order to re-charge its HP Air storage. It might in that condition have to submerge very quickly to avoid collision or attack, and so could find itself dived with no reserve of air for a further surfacing. Consequently, it is usual to provide enough HP Air for at least three full MBT blows before the submarine has to re-charge the system. With the requirement for twice the atmospheric pressure and three full tank blows, six times the volume of the MBTs has to be provided in the way of storage of air at atmospheric pressure. It is common practice to store the air in bottles at approximately 300 bar, so that if the HP air bottles are stowed in the MBTs about 2% of the tank volume is taken up with air storage.

To conserve HP Air it is usual to only partially blow the tanks when surfacing and then, when enough of the hull is above the water surface, to switch to a low pressure blowing system to bring the submarine to full surface condition. This alternative blowing system affords some safety back up in case HP Air is lost, because the boat can come to periscope depth dynamically and then use the Low Pressure Air (LP Air) blow taking in air through the ventilation mast. However, the LP blow system requires very large piping from the blower to the tanks to achieve a reasonable rate of water discharge. The HP Air system on the other hand needs relatively small piping. As the air eventually expands to a pressure not much higher than sea pressure there is a considerable differential pressure at the hull valves, which is sufficient to cause the air to reach Mach 1 velocity. In the main, displacement of water from the tanks takes place by progressive expansion of high pressure air within the tanks. Though the tanks need not be designed for full sea pressure, their structure has to be strong enough to take account of the possibility of some degree of over-pressure during the blowing sequence.

System arrangement

9.10   The storage of air usually takes the form of high pressure compressed air bottles, stowed in the MBTs external to the pressure hull or in free flood spaces. In some submarines, however, the bottles are wholly stowed inside the pressure hull and in all boats at least some of the bottles will be stowed inside the pressure hull. When stowed externally the bottles displace water in the MBTs and in design an allowance has to be made for their presence when calculating the reserve of buoyancy.

Since bottles are stowed in the tanks it might seem logical to provide short lengths of piping and dump the air direct into the tanks. Though some boats are fitted with such a system for emergencies, it is not easy to control and so is not adopted for normal blowing of the tanks. It is more usual to pipe the bottled air inboard to a central distribution panel from whence it can be directed to the tanks selectively and under control. This arrangement also allows the command to choose which of the several banks of bottles carried are to be used for a blow, maintaining other banks of bottles at full pressure. For that reason, the usual practice is to arrange a ring-main system, i.e. a circuit of piping along both sides of the boat and cross-connected at the ends and in the middle, to which each bank of bottles is connected. This ring-main can be divided by valves so that the banks of bottles can be isolated, thus providing options for charging some bottle banks while other banks remain available for use if needed (Figure 9.4).

While the foregoing is the long established way of arranging the HP Air system, it does involve a tortuous route for the air from the bottles through all the piping and valves before it eventually reaches the tanks, resulting in quite large pressure losses. It is also relevant that HP Air is a compressible fluid and therefore expands adiabatically through the system, causing increased velocities as it progresses through the system, which can also cause a rapid drop in temperature in the piping. A complex computation is required to determine the pressures throughout the system and the velocities and rates of flow that can be achieved from the

MBTs. The system could be simplified by more direct routes to the tanks whilst allowing the selectivity of the bottle banks by means of solenoid operated valves controlled electrically from the Control Room. However, most submariners like to have a manual fall back capability, which for HP Air, as for the hydraulic system, involves running high pressure piping to the Control Room or nearby, where an assembly of valves has to be located at which the valves can be opened manually to direct the flow of air to the selected tanks. The arrangement requires large amounts of piping throughout the boat, increasing the problems of bulkhead space for penetrations and occupying space in the vessel.

Also connected to the HP Air ring main, sited usually in the AMS or main machinery space, are the compressors for recharging the bottle banks. The process of compressing air to high pressures is inherently noisy and measures are taken to reduce the noise and isolate the compressors from the rest of the hull. The air compressors would normally be run on the surface to recharge the bottles that have been expended in the surfacing operation. However, on occasions they may be used whilst dived. Because of the other uses of air within the ship there is a tendency for the pressure to gradually build up inside the hull, which could be dangerous; a pressure build up increases the partial pressure of toxic gases in the atmosphere including oxygen and carbon dioxide and may also result in a hazardous situation when the vessel surfaces and opens its hatches. (It has been known in early submarine days for the man opening the hatch in the conning tower to be blown overboard because of the internal pressure build-up, and for many years it was common practice for another crew member to hold on to the legs of the man opening the hatch to ensure that he did not disappear overboard.)

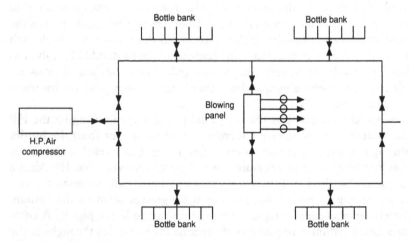

Fig. 9.4   Basic H.P. Air ring main system

The choice of number and sizing of the air compressors depends on the rate at which it is required to re-charge the air in the bottles. In older submarines which frequently surfaced, it was necessary to get the air back in the bottles as fast as possible so quite large compressers were required. For modern submarines which do not surface so often it is acceptable for the process of getting air back in the bottles to take more time. That can be done during the snorting period using the ventilation mast without the need to surface.

Other uses

9.11   There are other tanks on board including Q and D Tanks which may be used when the vessel is dived. In order to discharge water from them it is necessary to shut off an inboard vent to the tank and open the sea valve and then apply a HP Air blow to discharge the water to sea, stopping the blow sufficiently short so that an air bubble does not escape giving away the submarine's presence. Once the tank has been nearly emptied the hull valve is shut and the inboard vent opened to allow the air to vent back into the vessel, and this is one of the reasons for the build up of air pressure in the boat. It is important for the sequencing of the valves to be correct. It has been known on occasions that the inboard vent has been open with the hull valve to sea shut during the blow phase resulting in a considerable amount of sea water coming inboard.

If HP Air is used for emergency operation of control surface rams, additional large bore piping has to be provided as the emergency system is rate controlled using a lever on a valve located in the Control Room.

In the event of accidental flooding on or near the surface it is possible to isolate the watertight compartment involved and apply a salvage blow to inhibit the flow of water by raising the pressure to match sea pressure, say, two atmospheres. That action will stop or delay flooding sufficiently until further corrective action can be taken. If, however, the submarine is deep when it suffers a flood, the salvage blow would not be very useful, partly because the bulkheads may not be designed to withstand full sea pressure but also because the amount of air required to increase the pressure in a large compartment to the much higher sea pressure would be liable to exhaust all the bottles. The practice in the event of flooding at depth is: (a) to use dynamic lift and power to reduce depth as quickly as possible and (b) blow the MBTs to increase buoyancy.

At full diving depth the rate of discharge would be very low due to the much higher sea pressure. The total volume of air in the HP Air banks, if expanded to deep sea pressure, would be less than the MBT capacity and hence provide only a limited ability to counter the flooding action. If the boat could be made to start rising, the air that has been admitted to the tanks would begin to expand, increasing the counter flood effect. The aim in such a recovery action would be to discharge water from the

MBTs faster than the rate of flooding, but the tortuous nature of the usual HP Air Blow System reduces its effectiveness in that respect. It is because of those circumstances that some submarines are fitted with a short circuit system to dump the air from the bottle banks direct into the tanks in which they are stowed, thereby increasing the rate of blow considerably. Some care has to be taken with the design of such systems to avoid inadvertent operation in normal circumstances. Other possibilities have been considered for the emergency blow system, such as gas generators to provide a one-shot blow sufficient to get the boat to the surface in an emergency.

The HP Air System also provides a main source of pressurised air to top up other bottle systems in the vessel associated with pneumatic valves and the hydraulic system accumulators.

9.12   The air in the HP Air System, because of the way it is taken on board, is not particularly clean and does not really provide a sound substitute for breathable air on board the vessel. Therefore many submarines are fitted with an additional internal air system known as the built-in breathing system (BIBS) which stores air of a breathable quality in bottles at high pressure. The bottles are connected to piping throughout the submarine, with connectors for breathing masks so that in the event of smoke or fire or sometimes in an accident condition prior to escape, the crew members surviving can link into the system and have a breathable supply even when the boat atmosphere is no longer suitable.

In older submarines the HP Air System is also used as a means of discharging torpedoes. The torpedo to be discharged is loaded into a tube and the rear door shut and clipped. The tube is then filled with water to equalise with sea pressure and the bow door is opened ready for discharge. By opening the HP Air valve at the rear of the tube an air bubble is injected into the tube, forcing the torpedo out. The problem with the system is that the amount of air used has to be very carefully controlled so that no more air is injected into the tube than will fill it when expanded, otherwise a bubble of air will be released to the surface on discharge revealing the position of the submarine. The practice is to stop the blow before the torpedo is fully discharged and to open an inboard vent to special tanks so that sea water at pressure can flood back in to the tube and push the air back into an inboard air tank. Because of the difficulty of control it is unusual for that type of direct air discharge of weapons to be adopted in modern submarines. The system now commonly fitted uses pressurised sea water to discharge the torpedo from its tube using HP Air indirectly, by means of a ram or turbine: with that method the torpedo is actually ejected by a jet of water and so there is no risk of the air used escaping to sea.

As stated earlier, it has also been the practice in some designs to use

HP Air as a means of transferring water within the boat to correct the trim or to discharge water from compensating tanks. Such systems are noisy to operate and also could result in air leaking outboard and are not normally adopted in modern designs.

WATER SYSTEMS

9.13   In common with other vessels a submarine has several water distribution systems. The usual ones are those to remove the water that invariably collects in the bilges of a boat and has to be drained into a tank ready for discharge to sea, and those for the distribution of salt or fresh potable water for domestic uses and for cooling equipment. They are generally low pressure systems without any particular safety aspects, but do involve piping runs to and fro in the vessel. Some cooling water systems are direct to and from sea and operate at sea pressure. The integrity of those systems is highly important and means of rapid isolation in the event of failure are essential.

More specialised water systems in a submarine are those involved with the trim and compensation of the vessel as described earlier. The process involves taking on or discharge of water in order to correct for the weight/buoyancy variation, or for the movement of water fore and aft in the vessel in order to achieve longitudinal balance. In earlier designs of submarines these two modes of operation were recognised as distinct and it was common practice to provide a high pressure pump capable of discharging water to sea for compensation purposes (Figure 9.5) and, entirely separately, a trimming pump to move water fore and aft for balance purposes (Figure 9.6). For a period some designs combined the two into a trim and compensating pump with which the water was either discharged to sea or distributed to tanks in the vessel through the same

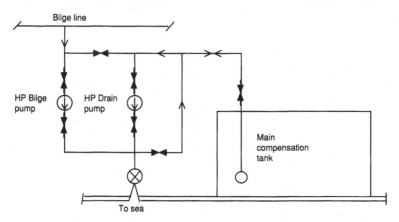

Fig. 9.5   Basic pump/flood compensation systems

pump and piping system, but that does lead to a large amount of piping on board which is subject to sea water pressure. As the diving depths of the submarines increased there was a heightened risk of failure of a piping system, resulting in rapid flooding, as it was under sea-water pressure. It is now usual practice to resort to something similar to the earlier systems in which the compensation system involves only short runs of high pressure piping with a high pressure pump and a separate low pressure distribution system between the trimming tanks, though this still involves a considerable amount of piping.

In some vessels a hovering system is fitted which also involves compensation water being moved in and out of the boat for the purpose of keeping a balance between weight and buoyancy whilst stationary. Such systems are usually specialised and they can call for high rates of flow, necessitating large pumps, though it can be arranged to use HP Air to back-up the system thus reducing the duty of the pumps by having a smaller differential pressure than would otherwise be the case.

9.14 Bilge systems and sewage systems take polluted water arising from various parts of the submarine and collect them in holding tanks so that they can be discharged overboard at a suitable time , whether for military reasons so as not to give the vessel's position away or, in peace time conditions, so as not to discharge polluted water to sea in unacceptable areas. In earlier designs it was usual to effect discharge by HP Air but it is now the practice to use pumps. The bilge pump may be common with the compensating pump and used as a back up, but a special pump capable of moving solids is used in the sewage system.

The provision of fresh potable water on a submarine depends upon the nature of its power plant. If it is of a diesel electric type, power supplies are limited and extensive use is made of sea water wherever fresh water is not absolutely necessary. Potable water is usually made from a distilling

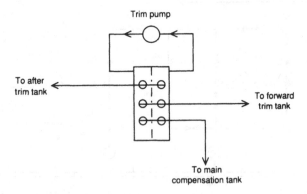

Fig. 9.6   Basic internal trim system [3 tanks]

system on board which, as a by-product, produces a highly concentrated brine solution which has to be provided with a holding tank for later discharge to sea.

Like any other vessel with a sizable crew a considerable amount of rubbish is generated on board and this has to be disposed of. Throwing the gash overboard is not practical and because discharge overboard is difficult whilst submerged, it is necessary to devise a special system known as a 'gash ejector' that will work whilst submerged. Rubbish is collected, bagged and weighted to sink, then placed in what is essentially a mini torpedo tube and discharged so that it goes to the bottom of the sea. As it is a potentially dangerous piece of equipment, because it could result in a large hole being open to sea if inadvertently operated, it is necessary to design an interlocking system so as to avoid mistakes in the opening of the outer and inner doors of the ejector. Care has also to be taken when filling the bags to ensure the removal of all air pockets and proper weighting, as there is always a danger that a bag could jam in the outer door so that when the inner door was opened on the next cycle the sea flushed the bag back into the vessel followed by a flood of water.

9.15 The machinery on board requires cooling which is provided either directly by sea water or indirectly by intercoolers to fresh water cooling systems, entailing taking on board sea water under pressure to the intercooler or condenser. If propulsion is by a steam turbine plant some of the piping is quite large, constituting potentially hazardous systems to have on board a submarine as they involve flexible elements introduced to isolate machinery from the pressure hull. The flexible elements are prone to fail and there have been incidents in which the submarines involved have experienced severe flooding and only just achieved the surface. It is in consequence, absolutely essential that such systems have rapid shutting valves directly on the pressure hull so that they can be isolated from sea in the event of failure. It is also significant that if a leak were to occur when a submarine was deep, the water spray that resulted would form an intense fog, filling the compartment and making it very difficult for the crew to identify the source of the failure and to isolate the failed system. In some vessels provision is made for the total closing of all hull valves as an initial safety measure in order to avoid flooding that could lead to disaster. The shutting off of all cooling water can, however, lead to consequential problems as much of the machinery, including that providing electrical power to systems, may also shut down. It is therefore considered that such an emergency action should only be taken as a last ditch attempt on the part of the command to prevent catastrophic flooding. Opportunity is accordingly afforded for the local watchkeepers to isolate the fault before switching everything off.

An implication of all these systems is that yet another large amount of piping has to be run to and fro throughout the boat requiring space. Not only that, there is a tendency to generate a hull penetration wherever a system needs to discharge overboard which can result in a large number of hull valves and penetrations in the pressure hull. There is much to be said for grouping systems so that only a limited number of outboard/inboard connections have to be made through the pressure hull, perhaps to inboard central storage tanks from which water can be taken off for the various applications. It might also be considered prudent that as many as possible of the cooling requirements should be met by having intercoolers fitted external to the main hull so that the sea water pressure is kept external to the hull rather than being brought inboard. Even the hull itself might be used as a cooling surface, because that would obviate any need for a penetration. In that context, it is a strange feature of submarines that trouble is taken to insulate the hull in order to maintain the temperature and comfortable conditions for living on board while at the same time bringing sea water inboard in order to cool those equipments that are higher in temperature. Systems which bring sea water inboard of the pressure hull represent a threat to the safety of the submarine and clearly every step must taken in their detailed design and by quality control in construction to ensure their integrity, reliability and maintainability at all times.

SYSTEMS FOR HYDROSTATIC CONTROL

9.16   The systems for hydrostatic control and the associated instruments and displays are brought together in a single panel in the Control Room, usually termed the Hull Systems Control Panel, so that the crew can readily carry out most of the operations described in this chapter by remote and centralised control, leaving only certain basic functions with essential safety purposes – like cottering the MBT vents when the submarine is to remain surfaced – to be carried out manually by designated watchkeepers.

The main hydrostatic functions carried out by centralised control are:-
Blowing or flooding main ballast tanks.
Operating trim and compensation tanks.
Redistributing contents between WRT, AIV and TOT tanks.
Blowing and flooding D tanks.
Flooding or blowing Q tanks.
Accurate instrumentation, particularly tank content gauges, is an important feature of the Hull Systems Control Panel.

ENVIRONMENTAL CONTROL SYSTEMS

9.17   For long periods the submarine is isolated from normal atmosphere and the crew on board have to be sustained in a totally enclosed environment. The atmosphere during a dive is continually in a deteriorating situation. The crew themselves are consuming some of the oxygen content of the air; at the same time they are adding to the carbon dioxide content of the air, so that if no corrective action were taken there would at some stage be insufficient oxygen for breathing and a hazardous level of carbon dioxide. The problem could be made more critical, as mentioned in the HP Air section, by increase in pressure of the atmosphere which would alter the partial pressures of the gases in the atmospheric mixture which could, under certain circumstances, become lethal. Not only does the crew reduce the breathable quality of the air but the many machines and equipments and operating devices on board release pollutants into the atmosphere which cannot escape. In the process, carbon monoxide can build up and several hydrocarbons can filter into the atmosphere from lubricating and diesel oils, while other equipments contain refrigerants which may leak to atmosphere. In themselves, those contaminants are not necessarily dangerous but they can result in dangerous gases being formed in the presence of sparks or fire. Some of the solvents used, particularly in modern adhesives, could cause dangerous pollution of the atmosphere as they slowly leach out. Some foaming agents for insulation materials are cyanide based and can slowly leach out many months after manufacture. As a sensible precautionary measure, restrictions are placed on the use of many materials in the construction of a submarine and to prevent the bringing on board of such materials which might add to these pollutants. Cooking on board releases fats into the atmosphere, together with smells; toilet and bathroom arrangements also result in smells which cannot be evacuated overboard as they would be in a surface ship. To counter all those hazards and dislikable factors a tightly controlled atmospheric regulating system is installed in the boat, which provides the necessary breathing gases and limits the build up of pollutants by removing as much as possible from the atmosphere.

9.18 The primary method of atmospheric control is to replace the consumed oxygen and remove the exhaled carbon dioxide from the crew. The provision of top-up oxygen to the atmosphere can be accomplished by the direct means of carrying stored gaseous oxygen in bottles under pressure. An alternative storage is by means of oxygen rich salts, a common form of which is the so-called oxygen candle. This is a mixture of iron and chlorate which, subjected to heat, forms ferric chloride and releases oxygen. Even when not used as a routine source of oxygen replenishment, candles are carried for emergency purposes and storage arranged in the designated escape compartments. The increasing adoption of air independent propulsion systems calls for a much greater oxygen storage, which may be in the form of liquid oxygen tanks or as High

Test Peroxide. The requirement for atmospheric oxygen can in those circumstances be added to the assessment of storage requirements. It should be recognised that none of these storage methods can be replenished during a patrol even whilst surfaced, so that the oxygen storage may become the limiting factor on the total submerged time during a patrol. Though compressed air is carried, it is not considered of breathable quality and as already described boats are fitted with special bottled air systems for use in emergency. In nuclear submarines in which electrical power is not limited, it is possible to provide a continuous source of oxygen replenishment by the installation of electrolysers to generate oxygen from distilled sea water. Great care has to be taken with the installation and maintenance of such plants as the by-product is hydrogen, which is discharged overboard through high integrity piping. The problem which must then be addressed is how to disperse the released gas so that it does not leave a detectable trail.

9.19   The provision of oxygen is essential to maintain breathing, but it is also important to limit the build-up of carbon dioxide in the atmosphere as this can lead to headaches, faintness and eventual loss of consciousness if its partial pressure rises to an unacceptable level. Some chemical systems are available for the removal of carbon dioxide by the process of combining with the carbon dioxide in the air to form carbonates, which results in a solid mixture. Such systems are limited in their capacity to remove carbon dioxide continuously from the atmosphere. If there is an adequate supply of electrical power, it is possible to remove carbon dioxide continuously from the air by means of what is known as a CO_2 Scrubber. This equipment makes use of the chemical reaction with Methyl Ethanol Amine (MEA), which will absorb CO_2 at higher temperatures and subsequently discharge it if cooled down, so that a regenerative cycle system can be maintained. By passing air flow from the ventilation system through a scrubber in which a spray of MEA is directed across the air flow, carbon dioxide in the atmosphere is absorbed; in a separate stage of the cycle the amine is cooled to release the carbon dioxide it has taken up, which can then be pumped overboard.

The main aim of the approach to atmospheric pollutants is trying to ensure that they do not arise in the first place, though inevitably some of the machinery on board releases pollutants into the atmosphere. It is necessary, in consequence, to equip the boat with a system by means of which the atmosphere is regularly monitored for pollutants and to provide a means of removing them. At the expense of oxygen the polluted air can be heated in a catalytic burner to breakdown hydrocarbons and carbon monoxide to CO_2 to be removed by the scrubber system. Fumes, fats and smells can be removed by a combination of passive and electrostatic filters and activated charcoal filters.

9.20 As well as achieving a tolerable atmosphere, it is also necessary to provide comfortable working and living conditions by controlling atmospheric temperature and humidity. There are many sources of wild heat within the vessel including the heat given off by each member of the crew and also sources of cooling, particularly in any areas where the hull has not been sufficiently insulated. There is a tendency for an increase in the moisture content of the air as humans exude moisture to the atmosphere, while cooking and other operations on board also result in an increase in moisture. It is important that an efficient ventilation system is fitted in the vessel which provides proper circulation of air to all parts of the vessel and avoids any dead air pockets. This is not easy to achieve and often in a newly designed submarine there are found to be parts of the vessel that need attention because they have insufficient supplies of air. The ventilation system is required to supply air at a suitable temperature to maintain comfortable conditions and a suitable relative humidity of approximately 50%. For that purpose all air in the boat has to be circulated through an air treatment plant which initially cools the air to remove the moisture and then heats it to a suitable temperature for return to the compartments. Although this is an energy wasteful process it is unavoidable as it is the only suitable way of removing excess moisture from the air. One of the problems with any ventilation system is the variable distribution of crew throughout the 24 hours as personnel go on and come off watch, are sleeping, eating or working. It is difficult both to design and to maintain a well-balanced ventilation system throughout the boat. Ventilation involves the provision of quite large trunking and piping to distribute the air and also large bulkhead penetrations to pass it between the compartments, which pose a threat to the watertight integrity within the submarine. In the context of initial design a significant space and weight allowance must be made for equipment and trunking associated with this sytem.

Fire prevention

9.21 Fire is an ever present danger in any ship and is a particularly serious hazard in a submerged submarine. A fundamental precaution is for the designer to eliminate or reduce as far as feasible the possibility of any fire hazard on board by the use of non-inflammable materials and also by taking care that any materials used when heated do not give off toxic gases. In the event of a fire, the first action is to isolate it and then treat it. Use of water for fire fighting is not an acceptable solution on board a submarine and therefore blanketing systems are necessary. A fire onboard a submarine will rapidly consume any oxygen and will result in smoke and pollutants of all sorts getting into the atmosphere and it might not be tenable for the crew to survive in the badly contaminated air. This is one reason for the provision of the built in breathing system (BIBS) of bottled breathable air. By means of face masks the crew can link in to the system at bayonet sockets located at intervals along the piping in each compartment. A man can move about the compartment by connecting to the nearest socket on the piping. In that way it is possible for the crew to take emergency action and hopefully be able to surface the boat without being overcome by fumes.

PROVISION FOR ESCAPE

9.22   Consideration of fire leads to the ultimate threat to the submarine, its vulnerability to flooding. If the boat were to lose control of buoyancy in deep water at some stage it would exceed the collapse depth of the pressure hull and then the submarine and its crew would inevitably be lost. However, flooding accidents may occur in shallower water where the distance to the bottom is less than the collapse depth of the hull and so, the hull remains intact but the boat is unable to return to the surface, with some or perhaps all of the crew still being alive on board. Such submarine disasters are, fortunately, rare but have a high public profile because, like mining accidents, they involve people being trapped or entombed alive in apparently inescapable situations.

A basic issue in the design of a submarine is the provision or otherwise of high strength bulkheads to isolate selected parts of the hull from the rest and so limit the extent of flooding throughout the hull in the event of an accident. To provide such strong bulkheads may enable at least some of the crew to survive in the non-flooded part of the hull. It is usual to consider the provision of at least one or perhaps two escape bulkheads on the philosophy that either the fore part or the after part of the hull will remain intact and the crew in that section will initially survive.

On the assumption that the pressure hull remains intact and a part of the vessel is free of flooding, the next design decision is how to extend, for as long as possible, crew survival on board, so providing opportunity for rescue. The crew need a breathable atmosphere but it may no longer be possible to operate the normal ventilation and atmospheric control due to the loss of power in the boat. For that reason, local means at either end of the submarine are accordingly provided in the form of oxygen candles and carbon dioxide chemical absorbent systems. For a limited time the BIBS can be used, although it is normally kept in reserve for the last phase of survival when some escape attempt is made from the submarine. There nevertheless remains the danger that atmospheric control may be poor and eventually the levels build up so that the crew are overcome. Even if a tolerable atmosphere can be sustained, the lowering of the temperature on board would make it extremely uncomfortable for the crew. The first essential is for the crew to survive long enough for rescue vessels to locate and reach the position of the bottomed submarine and be able to assist in escape procedures. It is usual to equip submarines with emergency buoys and signal ejectors in order to indicate the location of a stricken submarine and aid the detection process. Once a rescue operation has been mounted, the next problem is how to get the crew out of the submarine and back to the surface. Basically there are two options, the preferred one being the provision of an underwater escape vehicle which can go down to the depth of the sunken submarine, lock on to a hatch specially designed for the purpose and transfer the crew from inside the submarine into a holding compartment of the escape vehicle and then to bring members of the crew up in small numbers to the surface. That approach allows escape from practically any depth to which the submarine hull has survived. However, it does necessitate proper provision on the submarine of suitable escape hatches with the associated lock-on arrangements to suit the escape vehicle and on the availability of sufficient escape vehicles to enable one to reach the scene of the disaster wherever it may be. The locking on of an escape vehicle may not be an easy operation as it does not necessary follow that the submarine will be lying upright and level on the bottom . It may instead be at quite a severe pitch or roll angle, requiring the escape vehicle to make contact at a canted angle, which involves a very sophisticated vehicle of which the US Navy's highly capable Deep Submergence Rescue Vehicle is an outstanding example. If no such vehicle is available or suitable, the only other option is for the crew to escape from the submarine through the hatches and make a buoyant ascent to the surface.

Down to about one hundred feet depth it is possible for the crew to equalise the pressure in the escape compartment with sea pressure by flooding the compartment, which would enable them to open the hatch and escape to the surface using buoyant jackets. Early attempts to escape in that way provided an oxygen breathing unit for each member of the crew. However, oxygen above two atmospheres is lethal and consequently most of the people who attempted that type of escape from the *Thetis* died in the process. Since the pressure at one hundred feet depth is about four atmospheres, just a lungful of air at that pressure would actually last for some considerable length of time. It is possible to float freely to the surface as long as the crew are trained to expel the air in their lungs as they come up to the surface to avoid their lungs being blown up by the increase in volume as the pressure decreases on rising to the surface. Such free ascent methods were for many years a common method of training for escape of submarine crews, but have now been superceded by the hood system in which each crew member wears a hood so that his head is in a pocket of air as he rises to the surface. He still needs to compensate for the reducing air pressure as he rises to the surface by exhaling on the way. From a hundred feet it is possible for the crew to be under pressure for some little time whilst taking their turn to escape from the compartment and still not suffer any of the well known diver's problem of the 'bends', namely, the build up of gases in the bloodstream at the higher pressure which expand as a diver rises to the surface causing blockages in the circulation system. At depths greater than a hundred feet this effect becomes a real danger to the crew and it is necessary for them to avoid being under pressure for any longer than absolutely necessary. To that end it is undesirable to flood up the whole compartment to sea pressure, but to arrange for escape to take place within a specially designed escape tower with upper and lower hatches so that a small compartment is formed into which two or three men can get at one time. The lower hatch is shut and then the tower flooded rapidly, which then allows the upper hatch to be opened and the men to escape to the surface experiencing the minimal time under pressure. It remains a sensible precaution to have decompression facilities on board rescue vessels to treat any crew suffering from the bends. With tower escape, the time taken for the escape of all the crew will be extended as only a few can escape each time, but that has to be accepted

Some submarines are fitted with an escape capsule, which is carried on the submarine at the rear of the bridge fin. This is a spherical capsule in which, by very tight packing, the crew can get at one go, after which it is released to float to the surface. It represents a form of underwater lifeboat for the crew, but does involve a penalty in the design of the submarine because of the large additional appendage to the hull and so it is really only appropriate for vessels with a small crew.

It can be seen that provision for escape generally makes considerable space demands on the submarine. At the early stages of design the escape policy has to be decided on and appropriate allowances made in the sizing of the vessel.

ELECTRICAL SYSTEMS

9.23   It will already be apparent that a submarine would not be able to function without electrical power. Indeed, the provision of power supplies characterised by their high availability and integrity is vital both to the safe operation and the fighting capability of the vessel. In this section we concentrate attention on electrical systems fitted in conventional submarines. The systems in nuclear submarines have similar features, though the scale is altogether larger apart from the battery installation.

Typically, electrical power will be required so as to operate, or at least control, systems associated with the following applications:

Propulsion   Main propulsion system plus
                                       power for lubrication oil,
                                       hydraulics, cooling water and
                                       compressed air systems.

Control systems	Power to drive pumps, transmission and switches

Hotel Load	Domestic equipment, lighting, heating and ventilation.

Navigation and Communications
Weapons and Sensors   ⎤ Power supplies, signal processing
Emergency Functions   ⎦ and analysis and cooling

System equipment

9.24   The fundamental requirements for a submarine based power supply system are much the same as for any ship type and indeed most land based systems. Examples of the main requirements are to provide:

Sufficient capacity to meet demand
Supplies of the required quality
Reliable supplies
Satisfactory system protection arrangements
System operating flexibility

while being safe to operating staff.

Major systems elements

The main elements of the power system are:

The main generators
Electrical storage e.g. batteries
Main switchgear
Main power distribution equipment
System protection equipment
System control and surveillance

In addition, power conversion equipment is usually required in order to cater for variation in speed/power demands.

Features particular to submarines

9.25   The operating profile and the propulsion machinery of most present-day conventional submarines require the provision of a substantial storage capability for electrical power, which is usually achieved by means of lead acid batteries. To obtain a reasonable working voltage, a large number of cells in series are required (220 in series to to give a nominal voltage of 440V) and to achieve a high energy storage capability (long endurance without recharging) cells are connected in parallel, or a second main battery is provided, or both may be employed. This installation means that the main battery(s) dominate the design of the vessel. Conventional DC electric motor systems can presently be designed and built to operate much more quietly than AC machine systems, although that may no longer be the case in the future. Recognising that the main battery provides DC power and that stealth is a prime requirement of most submarines, it is understandable that their propulsion systems have been DC based almost universally until the present time. However, DC machinery and the associated control equipment tend to be large and heavy – very relevant factors in volume and/or weight critical vessels like submarines. Conventional DC equipment also tends to be relatively expensive. It is probable, therefore, that AC propulsion schemes, already becoming popular for commercial vessels especially the cruise ship market, will be developed to the stage at which they are fully suitable for submarine propulsion. The use of the diesel engine as the prime mover in submarines has been standard practice for many years and is likely to remain so for the foreseeable future, although it should be appreciated that new technologies under development may well change this in time. The most promising at the moment are air independent combustion methods, indirectly generating power or direct electrical energy generation, e.g. fuel cells.

Main features of propulsion system

9.26   Space considerations do not permit the adoption of dual main propulsion systems in submarines. With the adoption of a DC based electric propulsion system, therefore, it should not be too surprising that this is bound to influence significantly the design of the main power generation and distribution system. Indeed, it is not sensible to consider the design of the power system without parallel attention to the propulsion system.

Main generators

The rating of the main generators depends on:

The number of generators

The power demanded by the propulsion system when operating in diesel electric drive.

The required battery charging characteristics

The demands of the ship and weapon electrical loads

The specified growth allowance

In practice, two or three main generators are usually fitted, representing a compromise between the space needed and the required supply availability in the event of generator failure. In modern submarines, the demand for electrical power can be quite high, over one megawatt on occasions. To meet this demand using only two generator sets clearly requires that each individual set should be of a relatively large capacity. The use of a slow speed diesel with the power output demanded in modern vessels would result in a very large diesel generator, and so it is necessary to use high speed diesels. To avoid the use of a gearbox (to minimise weight/space) requires that the generator run at the speed of the diesel. A typical speed might be around 1200 rpm. There is, however, a fundamental limit to the power output of a conventional DC generator and its operating speed, expressed by the relationship:

$$\text{Power (kW)} \times \text{Speed (rpm)} < 1 \cdot 5 \times 10^6$$

which may result in the need for an alternative solution. Fortunately, a viable solution exists, namely to employ an AC generator, which does not have the same limitation on output, and to rectify the current to DC. An added benefit in weight and space is obtained since an AC machine is considerably lighter and smaller than its conventional DC equivalent, even allowing for the increase in weight due to the rectifiers, as the rectifying diodes can be fitted into the machine structure where they take up little space.

There are two features of the rectified AC solution that must be addressed at an early stage in the submarine design. The first regards the creation of structure borne noise by the generator. AC generators tend to cause noticeable line spectra at frequencies related to the speed – which is usually fixed. This feature may necessitate adoption of complicated mounting arrangements. The second feature relates to the quality of the DC supply feed to the various user circuits. The ripple content of the rectified AC will be much higher than that present in the output of a conventional DC generator. That may not matter, but with the abundance of electronic equipment present in modern power and propulsion systems, the acceptability of this feature must be confirmed by the designer. The magnitude of the ripple voltage can be reduced by employing a generator with a large number of phases, say, twelve instead of the usual three, so there is a possible solution to the difficulty.

Propulsion motors

9.27 Modern low noise propulsors operate at much lower rpm than those associated with power generation. The design of conventional DC motors operating at the required maximum rpm is unlikely to be a problem at the power levels of interest, although two motors per shaft may be needed. However, a significant problem relates to the range of rpm (and power output) required of the propulsion motors, the range being very large – virtually from zero to maximum. The problem is increased when the need to operate on the batteries is taken into account; this is because of the wide voltage range associated with batteries, depending on the state of charge. Solutions to the problems have typically centered on the use of multiple batteries and of two motors or armatures per shaft. Multiple batteries may be connected in parallel or in series thereby doubling the output voltage and increasing the shaft rpm and power output. Similarly, multiple armatures may be connected either in parallel or in series, which results in changes to the shaft output power/rpm. The variation for a given battery/armature connection scheme can be accomplished by means of field controllers for the motors. Another matter regarding the motors relates to their short term power rating. For long duration operation, the power output of the motors will clearly be matched to their supply generators (this is the motor continuous rating). Less obvious perhaps is that when the motors are supplied from the batteries, the power available may be much higher than that available from the main generators. That characteristic can be used to advantage in giving the submarine a high speed 'sprint' capability – provided, of course, that the motors have been designed for the higher power output. The operating scenario involved is necessarily a short time one; eventually the batteries will reach their minimum state of charge and the high rate of discharge process must be terminated. In consequence, the motors themselves need not be designed to provide such high powers continuously, which would make them very large, heavy and expensive. They need only be capable of providing the highest powers on a short time basis, in practice probably limited by motor and control equipment temperature rises.

The main batteries

9.28   We go on to examine the key features of the main battery. Although battery technology is steadily advancing, under the incentive of the potential electric car market, the new technologies have not yet reached the fully proven state needed to commit a major submarine design programme to their adoption. The well-tried and tested lead acid cell, albeit of much improved design and performance compared to the early versions, is almost universally used for the storage of the electrical energy need for submarine operation. As already described the main batteries are the heart of the submarine's power system, absolutely vital to the safety and operational effectiveness and so merit particular attention.

A significant factor in the design of the battery installation is the problem of connecting perhaps 250 or so cells in series/parallel combinations, or at least in series. There is the added problem of magnetic fields, for the large DC currents which the battery can deliver cause large magnetic fields, an undesirable feature for a military submarine! Then there is the problem of personnel safety whilst working within the battery compartment; lethal voltages exist between certain points within the cell interconnection structure. The designer's task, therefore, is to produce a battery copperwork layout scheme that minimises the generation of strong magnetic fields (by employing compensation techniques) and reduces the electrocution hazard by keeping points of high potential within the cell interconnection copperwork apart by at least the maximum spread capability of a man. The cells themselves need to be housed within a sump in order to contain spills of contents of a fractured cell case. Suitable pumping arrangements and sump level transducers also need to be provided.

Adequate personnel access to all cells is vital; cells require periodic topping up with distilled water, a source of which needs to be provided within the battery compartment; the cells also require to be checked for specific gravity using a hydrometer, and adequate headroom must be allowed to permit this instrument to be inserted into each cell. Those aspects represent an area where future development is likely to produce built-in continuous cell monitoring, the major difficulty being to develop a system that is intrinsically safe, bearing in mind the presence of hydrogen gas in the confined space of the battery tank. Hydrogen detection and removal facilities, usually a range of fans, must, of course, be provided. They are usually powered by the battery itself in order to ensure that power is always available when needed. The detailed design of the battery tank must ensure that no feature exists that might result in the gradual build-up of pockets of hydrogen. Facilities are provided to electrically isolate a 'sick' cell from the battery (not physical removal of the cell). The battery tank is fitted with heavy lifting facilities to permit the safe removal of all cells from the tank along a prepared exit route.

It is very likely that the cells will require cooling (water) in order to prevent overheating under the more onerous discharge conditions. Also, cell electrolyte agitation facilities (air bubbles) will probably be needed in order to ensure the maximum performance of the cells.

The battery tanks are usually sited at the bottom of the pressure hull and so arrangements must be made to prevent water or debris from falling into the tank through open access hatches.

Main power supply systems

9.29   We now discuss the issues which arise in providing power to meet the needs of the various services provided in a submarine, starting by examining the likely requirements and identifying problem areas.

Bulk power is provided as DC in most present-day submarines for the reasons already stated and the need is to establish the extent to which this DC power is suited to supply the various services.

Considering first the propulsion auxiliaries, most are motor driven mechanical equipments such as lub oil, fresh water, salt water, hydraulic oil and refrigeration pumps. The most direct design solution would be to employ standard DC motors and feed them directly off the main supply. It must be born in mind, however, that conventional DC machines capable of operation over the large voltage range associated with a battery on discharge are significantly heavier, more bulky and expensive than that workhorse of industry, the induction motor. Also, the maintenance workload associated with DC machines is far more than that for the induction motor, a point that must be kept in mind for a modern vessel in which the crew size will have been reduced to the minimum. Those considerations also largely apply to the motor control equipment (starters). The obvious course for the system designer to follow is to examine the feasibility and weight/size/cost advantage of adopting the induction motor whenever possible. To achieve the maximum cost advantage, it is necessary to use standard machines, i.e., those designed for 440V, 60 Hz operation – although the detailed design may be to military standards (e.g. shock resistance and noise reduction). Clearly, AC motors cannot be fed directly from the battery and therefore some form of power conversion will be needed, which may take the form of either a rotary convertor or an electronic (Thyristor) based converter. The features of the chosen converter must, of course, be allowed for in the overall feasibility assessment. For a submarine of reasonable size fitted with a large number of motors, it is most likely that the use of AC motors will be very attractive and the same may be true for smaller vessels.

It might not be possible, or prudent, to employ AC motors for all services. The question of system, hence equipment, reliability has to be addressed. This will be particularly so for systems associated with crew, weapon system or submarine safety. For such services reserve (alternative) equipment fed by DC machines directly off the battery may be provided.

Weapons and navigation equipments often require electrical supplies at 400 Hz. In those instances conversion equipment – either rotating or electronic – will be needed. However, the total power demanded at 400 Hz may be quite modest, which eases the conversion problem.

Control, indication and monitoring equipments are often based on 24V DC. The power demands are usually small and can easily be met using standard transformer rectifier units if AC supplies are available, or rotating convertors fed from the main DC supplies if AC power is unavailable. A further option in principle would be to bleed off 24V from the main batteries by connecting into the appropriate number of cells. However, this has the disadvantage of risking an unbalanced state of charge amongst the cells of the main battery, which is generally considered to be detrimental to the battery life.

Control and monitoring circuits essential to vessel or weapon safety are usually backed up by one or more small emergency batteries.

Discussion of system installation features

9.30   Power from the main generators, convertors and the main battery will be fed directly into a switchboard for subsequent distribution to the end users. It is usual practice to install most of the distribution control and monitoring equipment within switchboards.

The generators and the main battery are protected, e.g. against system short circuits and overloads, by the appropriate generator circuit breaker. Therefore, there is no protection between the generator terminals and this circuit breaker. For that reason, it is necessary to minimise the distance between generators and their switchboards in order to reduce the vulnerability to damage of the interconnecting cables. This aspect is of particular importance with the main battery switchboard, since the fault current, and hence potential for serious damage, that can be delivered by a battery will probably be much higher than that from a machine.

The current demanded from a battery may be very large and beyond the capacity of available electric power cables, which may require the use of copper busbars in the form of solid copper bars – possibly laminated in order to aid heat dissipation. It may not be a practical proposition to employ such bars on long or tortuous routes through the vessel, and so a site for the battery switchboard needs to be found in the vicinity of the battery. The problem presented by that need will be compounded if two battery switchboards are to be fitted – often the case in order to increase system operating flexibility and hence system availability.

In order to help increase electrical power supply availability – especially under damage conditions and bearing in mind that water and electricity do not mix well – the switchboards are normally designed to be air cooled by natural ventilation, as opposed to being water cooled in some manner. The method requires that the switchboard cabinet contains air inlets at the bottom and outlets at the top, and while they will usually be designed to resist water ingress to a degree – often by means of louvres and deflection flaps – there is clearly a limit to what can be achieved. Consequently, pipes containing water or any undesirable fluid (especially fluids under pressure) should not be run in the vicinity of the switchboard, a difficult precaution to observe in the confined space of a submarine.

A further point relates to the installation of the main power cables. Large power cables are not easily handled and that might lead to the running of several cables of manageable size in parallel. Although that alternative helps with the installation task, the penalties need to be recognised. Amongst these are:

Increased cost and weight

More cable entry positions in switchboards causing increased size and weight

Increased risk of cable damage owing to over-heating of cable bundles.

If the electrical distribution system features both normal and alternative supply arrangements, the associated cables should be kept separate so as to reduce the effects of a single fault event, for example fire, explosion or severance. Main supply cables from separate generators should also be run as far apart as possible, ideally well outboard to port and starboard for the same reason.

The siting of switchboards has to take into account the need for access for maintenance as well as for the installation task itself. The main power cables will probably enter the switchboard at the rear and adequate space to bend and insert these difficult cables must be provided. If the switchboard is to be mounted on shock or vibration mounts, adequate clearance between the unit and bulkheads, deckhead and adjacent equipment needs to be provided to allow for movement in service.

A further design aspect is the ventilation of the switchboard compartment. Heat generated within the switchboard escapes as wild heat into the compartment and clearly the ventilation system has to cater for this. Less obvious, perhaps, is that the most onerous duty for switchboards (and the maximum generation of wild heat) may occur under damage or fault conditions, very possibly in circumstances in which normal ventilation standards cannot be maintained. Accordingly, the detailed design of the various support systems and associated operating procedures need to be such that switchboard damage through overheating is avoided.

In order to reduce vulnerability to damage, especially system damage, it is usual to disperse equipments as far apart, fore and aft, port and starboard, as possible. However, the limited space available within a submarine will probably necessitate compromise compared to what is normally achieved on surface warships. Nevertheless, attention has to be concentrated on achieving the best possible layout of vital distribution equipments. Similarly, cables forming part of alternative distribution options should not be placed in the same cable runs.

Control equipment for main generators and motors should be sited as close to the associated machines as possible to reduce the vulnerability of the control cables to damage with consequential malfunction of the equipment. This requirement is especially important for motors, where a loss of the field current, for example, may result in a dangerous overspeed condition.

The presence of pressure-tight bulkheads causes special installation problems for electric cables of all types where they penetrate those boundaries. Suitable designs of penetration devices are available, but the installation difficulty lies in being sure that the arrangement permits adequate access for repair, test or replacement – a most difficult task where a large number of penetrations occur in a confined space.

Equipment repair/maintenance considerations

9.31 Compartment layout considerations have to take account of machinery maintenance and repair. For example, the provision of a removal route suitable for an entire major electrical machine is unlikely and so machines of that size have to be designed with *in situ* dismantling in mind. Meeting that requirement also calls for adequate lifting facilities as well as space in which to store temporarily large component parts. As an example, the heat exchanger of a large machine will probably be a single component of significant size and fitted on top of the main machine. It is possible that deckhead height limitations will prevent removal in a vertical direction and so the whole assembly will need to be designed to permit removal by sliding to one side, or possibly by splitting the unit and sliding it off from two sides.

The magnet of a large DC machine will usually be designed to split at the centreline to permit access to the armature for repair, removal or, more likely, *in situ* rewinding. Alternatively, the magnet may be of single piece construction designed to be slid along the machines longitudinal axis to expose the armature. This technique may also be adopted for AC machines where the stator will be a single unit surrounding the rotor. Clearly, adequate space has to be provided in the compartment to permit those activities and the detailed compartment layout should be such that the work-in-wake is as little as possible.

The main switchgear will probably not be immune from maintainability difficulties. The best utilisation of available space may require that circuit breakers are stacked within the switchboards. Large main circuit breakers are very heavy and bulky and heavy lifting tackle will be needed to permit unit removal; once again the provision of suitable facilities will not be straightforward in the confines of a crowded compartment.

9.32   A significant upkeep task is the maintenance of the diesel engines. The space constraints of a submarine may prohibit siting them in separate compartments. Modern high performance diesels for submarine application will invariably be supercharged; supercharged engines are very noisy, probably in excess of the damage limits of the human ear, which presents difficulties for maintenance as well as watchkeeping personnel who need to work in the compartment with one engine still running (simple earmuffs may not reduce the noise hazard over the whole relevant frequency range to safe levels). Indeed, this kind of problem motivates increased use of automation, including machinery surveillance systems.

A further point to be considered relates to maintenance or repair without impairing the operability of the vessel, which requires that flexibility and some redundancy be designed into the electrical supply network. The aim is to permit, for example, work to proceed on a major machine or switchboard without interruption to the important services required.

REVIEW
9.33   As we expressed at the outset of this chapter, although systems are of paramount importance to the operation of a submarine, they are

not an aspect that has to be considered in detail during concept formulation. The design of the systems will be undertaken by specialist engineers during the later design phases.

What is important in concept design is that the sizing of the vessel should make due allowance for the systems to be fitted. These allowances will be formulated as budget figures of weight and space which together with the performance requirements of the systems provide the framework within which the systems designers will work.

It is helpful at concept and feasibility stages if at least a line diagram indicating the configurations of each major system is produced, as this clarifies the selection of system type and goes some way to identifying the location of major components and system runs. If, for example, the concept design envisages a distributed hydraulic power pack system or a trim control system with a hard compensating tank and soft trimming tanks, then the budgeted allowances would be for those versions and not for alternative systems which the specialist designer might choose if no guidance were provided.

The dependency relationships for each of the systems are quite complex and may involve information which is not known or generated in the initial sizing of the submarine. It will be appreciated that each system is to some extent dependent on the size of the submarine and the relationship is sometimes explicit in the dependency or implicit in the demands, e.g. the ventilation and air conditioning system has to heat the total volume of air in the boat, while the hydraulic system has to provide control surface power which is related to size of vessel. Thus in the preliminary sizing process of the concept phase it is probably sufficient to make a volume allowance for systems and auxiliary machinery treated as a proportion of the hull volume, say, Volume required for Systems and Auxiliary Machinery = K x Hull Volume, K being a proportionality factor which can be determined for existing designs and modified if it is proposed to make significant changes in system design.

10 CONSIDERATION OF BUILDING
AND COSTS IN DESIGN

INTRODUCTION

10.1 Attention in previous chapters has been focused on the technical considerations in submarine design which bear on achieving the required performance. In the first chapter, however, we did point out that the designer has at all times to keep in mind that the submarine should be capable of being produced at an acceptable cost to the customer and be considered value for money, and also that the resources are available for the detailed design and build. Although it is conceivable that performance might be regarded as paramount, and consequently that any costs incurred either directly or to create the resources necessary for building have to be accepted, in most designs it is necessary to keep a strict balance between performance, cost and resources.

It is pertinent to consider the comparison between the cost of building a submarine and that of a corresponding commercial ship. On the basis of displacement tonnage there is a considerable difference in cost per tonne for the two types of vessel. The difference between the two calls for explanation.

One factor is the difference in the way in which the size of a submarine and a commercial ship is expressed. The gross tonnage of a commercial ship is in fact determined by its cargo or hold space and does not involve the weight of the ship itself. It is better to compare costs on the basis of the actual weight of the ship as constructed, termed its lightweight. Making that change, the cost per tonne for a commercial ship is higher but still significantly less than that of a submarine. Another factor is that the commercial ship is for the most part a large structural steel box within which only a small proportion of the volume has much in the way of fitting out. Even the fitted out parts are a relatively simple spaces with few systems compared with those in a submarine. The third and most important factor in the difference in costs is the extremely high density of packing in a submarine. We mentioned earlier that the submarine has an average density of unity compared to that for a surface ship of about 0.2 or 0.3, which indicates considerable congestion of fitting out within the submarine. Even those systems that are used in modern merchant ships, some of which are quite complex, are usually fitted in relatively open spaces. There is in consequence a striking difference in the experience of walking through a submarine and that of walking through the main operational spaces of a commercial ship. There is a temptation to speculate whether submarines would be cheaper to build if they were made larger and less congested, but although the instincts of many who have been involved in design and building submarines lead them believe that could be so, it is difficult to prove or demonstrate.

Before we discuss how the designer may influence the costs of the vessel and describe the resources required to build it, it is useful to consider briefly the way in which a submarine is built and also the costing methods that are available in determining how the breakdown of costs in a submarine is decided.

SEQUENCE OF BUILDING A SUBMARINE

10.2   Before fabrication can commence, all the basic materials and equipment that will be put together to form the submarine have to be purchased and stored. As regards the hull there are the many steel plates of various qualities, tolerances and thicknesses that are required to form the structure of the vessel. To this may be added standard structural sections where frames are not designed for fabricating. Next there are the many sizes and qualities and materials for piping and electrical cabling. For these there will be a myriad of components like pipe fittings, connectors, valves, electrical connectors and switches. Other important items are the major castings and forgings that will be required for propeller shafts, control surface stocks, etc. Some of the hull design may also call for special forged shapes to be built into the hull. Procuring these special forgings and castings can be a major constraint in the design and build programme and so they may have to be ordered early in the design process so as to be sure that they are available at the right time in construction. Thus design decisions may have to be made very early on and be committed for the rest of design at a stage when the designer might not wish to make such a commitment. Similar considerations can apply to many major items – machinery, equipments, electronics and so on – which may need a long lead in production at the manufacturers and, depending on the build sequence, have to be with the shipbuilder in time to be fitted at the right stage.

If the shipbuilder is a major industrial organisation it is possible that within the organisation, departments exist which can produce some of those items. If on the other hand the shipbuilder's role is that of assembling items supplied by sub-contractors he would have to adapt his build programme to suit the quoted delivery dates for the served-in equipment.

The ordering in of materials and equipment is a major exercise requiring a substantial organisation in the design offices and ordering departments in order that the process can be executed efficiently and to the right time schedule. The use of CAD/CAM systems can considerably help in this process as it is possible to extract from the data in the computer information on all the items that are required.

Before assembly and fabrication of the material start, the shipbuilder's quality assurance organisation plays its part in ensuring that all the materials meet specification requirements.

Steel work

10.3   The steel plates that have been supplied will be cut to the required shapes for the various parts of the hull and internal structure. The CAD/CAM systems also help in this process by producing numerically controlled cutting programs and conserving materials by the careful nesting of small pieces to make maximum utilisation of the plating.

For many of the internal structural elements of a submarine the flat pieces are assembled and welded together to form box-like structures, but for much of the pressure hull there is a requirement to form the plates and sections by rolling them to the curvature of the hull. The choice of thickness and tensile qualities of the material in the design determines the amount of effort and the types of machines required for the forming process. Additionally, if the design calls for domes or spherical transitions in the pressure hull, the plates have to be given double curvature in forming rather than the single curvature of the cylindrical sections. For very large domes this requires the pressing in dies of plate elements that can be welded together to form the dome. For smaller domes it may be possible to spin the whole of the dome out of a single sheet, which involves the hot spinning and subsequent heat treatment of the plating once it has been formed.

The pressure hull frames, usually of T-bar section, may be fabricated or supplied as extrusions. The tables of fabricated frames are rolled to shape, but the entire extruded frames have to be pressed to shape.

Fabrication of the main hull structure

10.4   The usual procedure is to assemble the plating and frames into lengths of two to three metres of the hull and these are known as hoops. Some of the hoops will be cylindrical, others will have slight conicality and others near the stern may have quite marked conical shaping to them (Figure 10.1). Controlled conditions are required for the welding, particularly with high tensile steels, and it is now common practice to have some form of very large mandrel on which the whole hoop section can be rotated to allow down-hand welding on both internal and external welds of the hull. With all main structure, the requirements are for welding from both sides of the plating to form through welds and so care in design must be taken to ensure that there is no geometric configuration in which only single sided welding could be accomplished because of the access (Figure 10.2). The use of jigs and mandrels for the assembly of the pressure hull hoops helps to control the circularity and dimensions which are a requirement in the design for pressure hull strength and also enables the shipbuilder to control dimensions so that adjacent hoops can subsequently be aligned for joining together.

The next stage of fabrication is to join the cylindrical hoops together with cones and domes and some of the external structure to form a few major sub-assemblies of the hull say, three or four major assemblies (Figure 10.3). At the appropriate stage, while access is still good, the simpler sub-assemblies of internal tank structure are introduced into the hoops and welded into place. During sub-assembly fabrication quality assurance is applied using non-destructive testing of the welding. Tank pressure tests can also be commenced once the structural work has been completed.

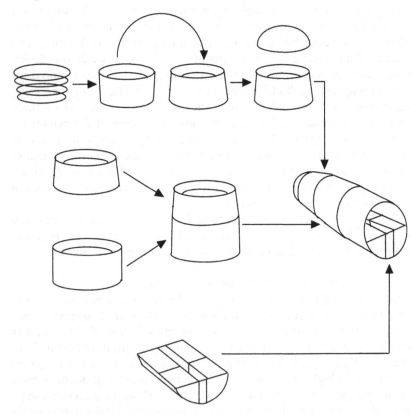

Fig. 10.1 Fabrication of hoops into berth unit 1

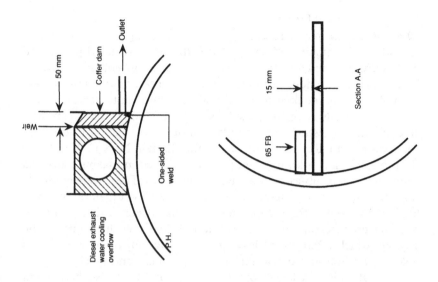

Fig. 10.2 Difficult accesses

Internal fitting out

10.5   While the main structural fabrication is taking place and in parallel with it, machinery and equipment with their associated piping and
cabling are collected together and assembled into convenient sub-assemblies. This can be done separately in clean workshop conditions away
from the slip or dock where the hull structure units are being put
together. One advantage of the sub-assembly method is that a considerable amount of testing can take place prior to installation in the hull.
This was a method used extensively by the German Navy in World War
II, in which many assembly groups were involved in making parts and
components and sub-assemblies, termed modules. These were gradually
brought together into fewer major sub-assembly organisations and
finally into two and three shipbuilders for final assembly. Previous shipbuilding practice was to complete the hull and then fit it out on the
slipway, which often meant leaving, or cutting out, quite large access
holes in the hull in order to introduce equipment and machinery into
the hull after it had been completed structurally. It also had a bad
feature in that pipe and cable runs were not determined until the time of
actually installing them in the boat and there was a tendency for the first
trade on board to take the best runs, after which the other runs of cabling
and piping had to make do with what was left in the way of space. In an
attempt to solve some of those problems, full-scale mock-ups of difficult
spaces in new submarine designs were built in wood to allow the details
of locations of seatings, equipments, pipe and cable runs to be worked
out in advance and to enable any unsatisfactory areas to be readily
removed and replaced in more appropriate ways. Mocking up also
helped because it conveyed the full-scale effect of the intended arrangement and could aid determination of access and removal routes within
the hull. More recently, submarine designs have been developed using
fifth scale models rather than full-scale mock-ups and now the use of
computer aided drafting has enabled quite intricate details to be investigated with a computer representation of the hull and much of its contents.

Returning to the outfitting of submarines, each of the fully tested
sub-assemblies or modules of machinery, systems and equipments which
may be mounted on rafts or parts of the hull seating structure, can be
installed in the appropriate structural sub-assembly while still separate
from its adjacent sub-assemblies, making the whole process of fitting out
much easier. Where the sub-assemblies or modules have inter-
connections within their section of the hull these can be completed,
finalised and tested before adjacent sections are joined together. For
those parts which connect to other equipment in other sections, it is necessary to leave tails on the pipes and cables and to make those joins at a
later stage.

Closing-out the hull

10.6   Ideally, the final joining of the four or five major hull units on the slipway or in dock should be at a near enough completely fitted out stage before the final hull welds are made. (Figure 10.3) The final stage, after the hull has been closed out, is to join together the inter-connecting piping and cables between the various sections, which should be a relatively small amount of work. Usually, there is quite a lot of detailed work necessary in way of the joins, particularly where sub-assemblies have been left unfinished so as to give clear access to the main hull joints for welding. Using this building sequence the submarine when launched from a

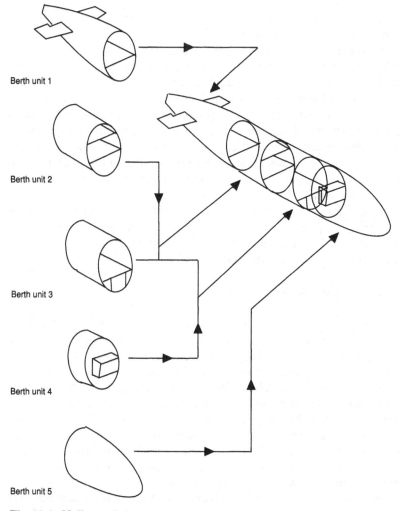

Berth unit 1

Berth unit 2

Berth unit 3

Berth unit 4

Berth unit 5

Fig. 10.3 Hull completion

slip or from a dock would be about 90% complete. Items often not yet fitted at that stage are control surfaces and propeller and some external equipment such as sonar arrays which are fitted at a subsequent docking, though in the case of launching from a dock even some of those items might already be fitted, so that the submarine is virtually complete. Subsequently, a considerable period of time is required to round off the outfitting, for setting to work of machinery and to test and tune systems. It is then necessary to carry out a series of basin dives to check the hull and its contents for submerged operations before the submarine is ready to go on sea trials.

COSTING

10.7   In this section discussion is centered on the costing process in a submarine design. The costing can be approached in many ways, none of which is particularly satisfactory as things stand at present.

It is important to distinguish between cost and price. Essentially the price of a submarine is the amount a shipbuilder or contractor is prepared to tender to build the vessel to specification. To an important extent the price will depend upon what one might call the politics of the situation which may involve aspects such as: the question of how many boats are planned to be built and at what frequency; the degree of competition; and the resources, facilities and expertise of the shipbuilder. In consequence, the price of a submarine can vary quite widely, even with exactly the same design specification. Cost, on the other hand, can be regarded, for our purposes, as an inherent property of a submarine, determined by the individual nature of its contents. Costing is the process by which the totality of individual costs is assessed, in our case by the designer during the early stages of the design.

Costing by weight groups

10.8   In cost estimating it is common practice to use the weight group breakdown which, as discussed in an earlier chapter, was devised for the purposes of weight estimation. By determining an average cost per tonne of the major groups at 1 digit level, like structure, machinery, outfitting, etc, a rough estimate can be obtained in early concept studies for the design of a new vessel. The quality of the weight data depends upon the information from previous vessels and its applicability to the new design depends on there being no large changes or innovations. A more refined costing can subsequently be obtained when the groups can be broken down into sub-groups at 2 or 3 digit level, for which average costs appropriate to that level can be assigned, e.g. main pressure hull structure in high quality steel will have a higher cost per tonne than secondary or external structures in ordinary quality steel. Clearly some care has to be taken if, for instance, a decision had been made to use glass reinforced plastic for external casing and fairing structure where previous designs did not use that form of construction. Weight-based cost estimating is a convenient way for costing because, having been used for many years by cost estimators, that is where their experience lies and, moreover, the historical data base for costs exists only in weight group terms. Nevertheless, it will not cater for unconventional geometries or unusual materials and it certainly does not fully reflect the functions of the submarine. For instance, is the cost of the hull structure a function of the diving depth adopted in the design, or of the speed and performance required of the submarine causing increase in size to accommodate the propulsion systems? (The point here is that even with similar technology, if the balance between submarine functions changed substantially attributions of cost to weight could be distorted.)

Cost returns

In some instances it may be possible for the designer to obtain cost returns from a shipbuilder for a previous vessel, though they are not likely to be available if competitive tendering for the design and build is involved. Even if they are available they may not be of much help to the designer in assessing cost against function, as the costing data will probably be in the form of the cost of bought-in materials, together with the costs of each trade or department. The data will not be directly related either to weight groups or the functional aspects of the design. A further source of difficulty related to a shipbuilder's cost returns may be the different ways in which the overheads, margins and profits are distributed across the workforce and administrative support departments. It can become very difficult to identify the proper cost of, say, the hydraulic system in a submarine and how cost might be reduced by a revision of the design of such a system.

Functional costing

10.11   Ideally, for design purposes, it would be helpful if costing could be related directly to the functional performance parameters of the design. Attribution in that way would enable the designer to identify to the naval users the cost consequences of the requirements which they were considering. However, the objective of functional costing, while desirable, is difficult to meet because it is almost impossible to obtain a single valued function/cost relationship. For example, the cost of a bulkhead can be quite easily assessed but how does one relate that cost to the several functions which the bulkhead performs in the submarine. It performs a function of supporting the main pressure hull against diving depth; it also acts as a boundary in separating compartments of the vessel; it provides support to decks and equipments; and it may also act as an escape bulkhead. It is possible that if the design was investigated without one of those functions there would be little change in the outcome, e.g. the necessary structure for, say, escape purposes makes the structure totally adequate for its other functions. Experience has shown that a customer faced with too high a cost may decide to relax a requirement which he regrets giving up, only to find that the designer is obliged to tell him that no saving accrues from this major concession. In principle, at the concept and feasibility study stage, it is possible to run and re-run designs changing individual parameters one at a time and determining from the costing assessment the difference in costs brought about by the changes. However, the non-linearity and inter-dependence of sub-systems can give rise to anomalous variations in costs in that parametric survey. Particular combinations of parametric changes may bring about a more beneficial change in cost than single valued function change, which could be described as a synergistic effect.

A suggested breakdown for design guidance purposes

10.12   For the purpose of identifying associated costs inherent in the process of design the following breakdown is suggested as a framework of reference within which to discuss the problem.

(a) **Identifiable costs** These are the items of bought-in materials and supplied equipments (including weapons) which can be readily identified as individual cost drivers in the design. Some care has to be exercised though, for while at the time of design it will usually be possible to obtain raw material costs per ton, by the time the material is actually bought-in there may have been a significant shift in costs.

(b) **Costs inherent in fabrication and assembly of the submarine** As discussed these costs are capable of being assessed but they may differ for different shipbuilders owing to the facilities and the resources they have for fabrication or manufacture as compared with bought-in sub-assemblies.

(c) **The capital costs and overheads** For a submarine to be built there will be a capital cost element in the facilities, machines and land, etc, needed to construct, assemble, install the contents and test the vessel.

(d) **Costs of applied standards, quality assurance and testing** specifications that are implicit or explicit in the design.

(e) **Costs of any research and development** necessary for design and building.

(f) **Profit margin.**

DESIGN INFLUENCE ON COST AND BUILDING

Using the headings suggested above we can now go on to discuss how the designer might influence the cost of a submarine.

Identifiable costs

10.13   Although submarines inherently have special demands and rigorous safety standards, the designer should nevertheless keep in mind that avoidance of high cost materials is a valid design objective where this is possible. For the pressure hull, if the design is volume controlled, it is conceivable that the hull could be constructed of a lower grade, cheaper steel with easier fabrication, accepting that it will then be heavier. Another way in which the designer could influence the cost of materials is to use materials which are in quantity production to standards of thickness, dimension and tolerancing which are normal within the industry. Similarly, standard materials for piping and cables, provided they are suitable for the submarine environment, should be adopted wherever possible. The design should also make use of readily available fittings, couplings, screws, bolts, valves, switches, etc, which can be bought from quantity production manufacturers. There has been a number of cases where unnecessary deviations from normal industrial standards have resulted in high cost of some of the bought-in materials because a low production run had to be made for the order. Nearly all warship equipment, whether for submarines or surface ships, tends to be several times the cost of commercially available equipment, primarily because of the shock standards required. In many instances it might be possible either by the siting of equipment or the design of seatings to avoid that cost element. Although the foregoing might be considered good advice , it can happen that the severe conditions of the submarine environment preclude being able to take advantage of the opportunities afforded by the use of commercial equipment. With some of the equipment, the designer may not have much option because it has been specially developed for submarine application, which is the case for weapons and sensors, or its use has been stipulated by the customer. However, there may still be cases where the designer can choose between alternative equipments with perhaps some lowering of cost. It is a common belief that the cost of the major weapons and sensor equipments dominates the cost of a submarine, but that is not entirely true.

Fabrication and assembly

10.14   A large proportion of the cost of a submarine is involved in the fabrication and fitting out processes. This is an area which the designer can considerably influence by his choices within the framework of the design. The watchword for the designer must be: wherever possible, keep it simple. If the design allows, the pressure hull should be a constant section cylinder using thicker plating and fewer larger frames with wider spacing, which would considerably simplify the fabrication process, reducing welding and allowing a longer production run with the same set of jigs for the rolling of frames, shell plating and assembly of the hoops. In contrast, varying hull diameter with many coned sections would necessitate every hoop unit being different with high fabrication costs. The wider frame spacing allows easier access between frames and reduces the amount of cutting and special structural design around penetrations. Wherever possible the design of pressure hull structure should avoid geometries entailing awkward joints which make for difficult assembly and welding; in particular the juxtaposition and geometry of the frames should avoid the need for single sided welding. Although it is usual to have end closures in the form of domes, which require double curvature pressing and welding of petals to achieve required shape, the avoidance of large changes in diameter with just a few coned transitions of gentle slope reduces the amount of complicated fabrication necessary.

For the same reasons the design of systems should be kept as simple as possible. Though safety and failure mode analysis may lead to the need for duplication of systems, that outcome should be avoided as far as possible because too much in the way of duplication not only increases initial fabrication but also considerably adds to the quality assurance testing and subsequent maintenance within the boat. It is also advisable to keep down the amount of pipework and cabling that run end to end throughout the submarine. Wherever possible the main runs throughout the boat should be just those for the power distribution systems and other systems should be localised to one compartment. That would maximise the use of pre-tested modules which can be pre-assembled and installed in the main hull sections with relatively few connections and final joining.

In the layouts and arrangements of the vessel the aim should be kept in mind of trying to create natural breaks along the hull where there are minimal cross-connections, so that they will be suitable places for the hull units to be joined together. Reasonable clearances in those areas are conducive to completing the final closure hull welding. Another factor which has to be borne in mind when designing submarines is that, though the structure may be adequate when the vessel is complete, during the building phases some of the mutual support between sections will be missing and so attention should be paid to the strength and stiffness of the individual sections which will be built as separate units. Yet another aspect is the necessity or otherwise for cutting large access holes to introduce machinery and equipment at a later stage in building, which can also lead to temporary weakness of the hull structure. Consideration of the modular concept in arrangement and of the removal routes for major components can be worthwhile in deciding how to make provision for possible changes, either during build or later life. The longer the period of time between concept design and completion of a new submarine the more probable it is that pressure will mount for changes to the design, purely due to the passage of time. It is important that the designer should resist that pressure as changes late on could cause dislocation to the building programme with consequent delay and severe cost increase.

Capital costs and facilities

10.15 During the design stage it is always relevant to keep in mind what industrial capability is available for the construction of the new submarine. In large countries it is probable that this task will be well within the capacity of the industry, but if vessels are to be designed in one country and built in another then the capability of the latter is inescapably a factor to be taken into account in the design process. If the performance of the submarine is paramount it may be necessary to expend quite a high capital investment in creating facilities to build it. High yield materials such as special alloy steels will be advantageous in the design of the vessel, but they are harder to form than ordinary steels and require heat treatment, temperature control and special welding techniques. If the design allowed the use of simpler materials without those necessities, considerable savings in capital and facility costs could accrue. Very large forgings and castings in special materials can result not only in long lead items but also in very limited availability of sources of supply, which cause increase in costs both because of the special nature of the forming and of the limited competition in obtaining a supplier. The casting of special bronzes for main sea-water valves can pose problems in requiring elaborate casting facilities in order to produce good quality, sound castings meeting the stringent quality assurance specifications necessary for such important components of the pressure hull envelope.

Costs of standards, quality assurance and testing

10.16 Some mention has already been made of material standards. Another factor to be considered in that context is the setting of manufacturing tolerances such as the circularity of the hull fabrication and, more generally, of the impact of quality assurance. Difficulty of complying with quality assurance requirements for special materials leads to applications for waiver of the requirements. If waivers are allowed, the implication is that the requirements have been set too high. The more elaborate the quality assurance procedure the greater can be the difficulty in achieving reasonable consistency in the assurance of quality during building and subsequently in testing and maintenance in service. That is not meant to be a plea for the avoidance of quality assurance but rather for keeping within reasonable bounds the need for it to be applied in the first instance.

The effect on design of research and development

10.17 It almost goes without saying that any advances in design will require research and development in order to develop test and analysis tools for use in the new configurations or fabrication methods. In that regard the designer plays a double role, one of which is to conduct concept studies involving possible innovations in configuration or equipment, assessing the consequent advantages in performance or cost that might be achieved. In a simple example, if it were possible to reduce the resistance of a submarine by 50% or alternatively improve the overall propulsive efficiency by 50%, that would give unquestionable improvements either in performance or in reduced size and cost.for the same performance. Other changes would be likely to have more subtle or less dramatic effects, but nevertheless could be considered worthwhile advances in the design. The achievement of such improvements would require research to investigate possible ways of improving performance and then the development of a physical realisation before they could be taken as reasonably established measures for inclusion in the design. The designer's role in that respect is in prompting and initiating candidates for future research and development.

However, the designer has another and quite different role in which he has to adopt a rather hard-headed attitude. He must really try to keep design within the realms that existing research and development have shown to be achievable. For example, an entire change in hydrodynamic configuration may offer attractive performance improvements, but what is significant are the unknowns of its behaviour in propulsion and control terms and how long would it take to find out and so be assured that the change could safely be incorporated in the design. Similarly, departures from the standard pressure hull configuration to enable, say, better internal layouts might cause initially unidentified structural strength or fabrication difficulties which only come to light at a later stage in the build programme. For the main pressure hull the adoption of an even higher yield steel or a new material such as titanium or GRP would pose new problems in the forming and connecting of these materials together and possibly longer term problems such as fatigue under cyclic stress. If the design were to make it imperative to adopt such special features then the cost and time of the research and development would have to be taken into account in adopting them. That is not to say that the avoidance of innovations is advisable in design, but rather that it is only realistic to recognise the risks involved. It is not just that the proving of new technology will take time, but also that it will require a change in the expertise of shipbuilding staff and therefore a training programme to impart the new skills.

Thus the designer needs to behave in a somewhat schizophrenic manner, sometimes as an innovative and imaginative person willing and able to stimulate research for the future and at other times as a very practical person in keeping his design strictly within known technology. Some design project managers adopt a rule of allowing, say, only around 10%

innovation in any new design so as to be sure that it can be produced within cost and programme intentions.

Profit margin

10.18 The designer has little control over this factor but clearly a shipbuilding firm has little interest in bidding for the construction of a submarine without a reasonable return on investment.

It would be helpful in assessing costs if the profit element could be kept separate and explicit, but it could be inextricably distributed over many cost elements, which makes for difficulties in assessing true cost. At one time warship building was contracted on a cost plus basis, which effectively gave the builder a guaranteed profit whatever the eventual costs of building turned out to be, giving little incentive to the builder to control or reduce costs. The opposite contractual arrangement is termed fixed price, where the vessel is built for an agreed total sum, determined at or near the outset. There is with that approach a greater incentive for the builder to reduce costs, but little gain to the customer unless the agreed figure is the result of competitive tendering or hard bargaining before the contract is placed. If this were not achieved the builder could make exorbitant profits. If, on the other hand, a really hard bargain is struck, considerable work could be caused for the design project as any real or apparent change or deviation from the original specification could be interpreted by the builder as an appropriate opportunity to claim additional costs.

SUMMARY

10.19 In this chapter we have tried to indicate how the designer can influence buildability and cost in the detailed design of a submarine, because some of the details are determined in the earlier design stages of concept and feasibility when the designer will be well advised to maintain an awareness of the practicability and to cost implications of his decisions.

Given the assumption that other aspects are equal, the cost of a submarine is largely determined by its size and complexity, which has led to the over simplified belief that by limiting the size of a design the cost will be controlled. However, if that reasoning is associated with no relaxation in the operational requirements of the design, the end result will be a more complex and highly congested vessel which could turn out to be difficult to build and maintain. Even if there were some reduction in building cost, which is doubtful, through life costs could be increased, negating the purpose of the exercise.

To achieve cost reduction by size limitation the designer must address those factors which influence size. From what has been discussed the dominant factors for a submarine are the vehicle performance parameters, of which maximum speed is most dominant, followed by the demands for weapon stowage and the size of the crew. Although the first of these can be manipulated in a conventionally powered submarine there will be little scope in a nuclear submarine because of the fixed reactor plant.

11 GENERATING A CONCEPT DESIGN

INTRODUCTION

11.1   It can be recognised that throughout the book we have placed emphasis on conceptual formulation in submarine design and on the high level of interaction between the sub-systems. It must nevertheless be appreciated that the successful outcome of a submarine design is ultimately achieved by extensive detailed study of every aspect of the design. It is only by detailed studies that the multitude of interaction effects can be identified and resolved. However, such studies have to be conducted within a framework which adequately defines the total design and that is the purpose of the Concept Design phase. The objective is to determine a size and weight plus geometric configuration within which the detailed studies (often conducted by specialist designers) can take place. For example, there is little point in conducting a detailed design of pressure hull structure if the envisaged configuration of the hull precludes achievement of a weight/buoyancy and longitudinal balance. Similarly, there is little point in a detailed investigation of the weapon compartment layout and tankage if its location within the hull is inconsistent with the fore-end configuration.

11.2   Consequently, the designer has to be able to determine the broad design characteristics having an awareness of the details that must follow. He is unlikely to have much information on those details, particularly in an innovative design, but he must endeavour to ensure that the subsequent stages are capable of achievement. It is possible that decisions may have to be made on the basis of comparative concept studies which could be rendered invalid if subsequently configuration problems proved impossible to resolve. If the design incorporated many existing equipments and standards then information from previous built designs would provide sound guidance on achievability. If the design were highly innovative or incorporated much new equipment or plant it could be necessary to conduct independent studies of local configurations to obtain a 'feel' for what is possible and necessary to achieve final design success. Such studies should not be confused with the main stream of the design process as they would not include many of the interactive considerations brought to bear in designing within the concept design framework.

CONCEPT STUDIES

11.3   The basis for initiating concept studies for a new design can be described under a number of sub-headings.

(a) The role of the vessel

This arises from operational studies of the scenario in which the submarine is to function. The studies are undertaken by the operational staff supported by operational analysis, and identify where and what the submarine is expected to do. Thus a military submarine may be required to act in the 'barrier' role, that is, to take up station in an area where an opposing force may need to transit and to be capable of detecting, identifying and tracking such vessels. The submarine may then be required to intercept the target or alternatively communicate its information to other intercepting forces. A commercial vessel may be required to detect a seabed pipe line, follow its route and carry out some form of inspection.

It is important that the designer should be fully informed of the intended role for otherwise the resulting design may not meet the expectations of the operators. If that was the likely outcome, the operators should be fully acquainted with the assessed shortfall in performance as it would necessitate alteration in the assumptions of the operational studies.

Whilst the new submarine has been discussed in the singular, this exploratory stage may also involve a debate as to the number of submarines required and their individual capabilities, e.g. a few, high capability, high cost vessels versus more, lower capability, lower cost vessels.

(b) Equipment

From the role of the vessel it should be possible to identify the operational equipment required to be mounted in or deployed by the submarine. There will be various types of sensors, communications and navigation systems and types of weapons with the number of reloads, which will enable the submarine to perform its anticipated task when on station. This set of equipments together with the associated operators (manning) constitute the 'Payload' for the design.

It is possible that there may be other selections of equipment as alternative ways of performing the operational task and this will lead to alternative concept studies of the submarine.

(c) Characteristics

The role will also lead to the identification of the 'Vehicle characteristics' required of the submarine. These are maximum speed, transit times from base to operating area, submerged endurance, indiscretion ratio, vulnerability and own noise standards, patrol endurance and diving depth. Again there will be the possibility of alternative vehicle characteristic packages.

Some of the values of parameters used in the concept sizing process can be deduced from the general requirements set by the operational staff.

Duration of patrol

11.4   The duration of a patrol in days will be the time spent on station plus the transit time taken for the submarine to make passage to and from base to the patrol area.

The transit time can be varied by changing the assumed transit speed, but it should be recognised that increased transit speed requires more power and hence greater fuel consumption. This penalty arises because power is proportional to the cube of speed whereas the time of transit varies only linearly with speed. Another consideration, which will be treated in more detail later, is that, if the transit is to be accomplished by a mixture of snorting and submerged running, an increase in submerged speed results in a higher indiscretion ratio. Hence the maximum transit speed may be fixed by the maximum snorting speed and the allowable indiscretion ratio. For conventional diesel electric boats with long runs from base to operational area there may be a case for conducting the majority of the transit on the surface and only adopting a covert transit nearer the operational area. That approach would raise the question of whether the hull should be configured for low surface resistance with an operational penalty on submerged performance or whether a submerged performance configuration should be retained at the expense of higher fuel consumption when surface running.

The time on station is not necessarily a variable of free choice. If it is required to maintain one submarine on station for a prolonged period then the time on station will depend on the time it takes for the next boat to relieve the submarine already on station. If we assume a two boat system then one boat must remain on station for the time it takes the other boat to leave station, return to base, carry out a base maintenance period (usually fixed), re-store and transit back out to station. A typical base maintenance period is 25 days and if the transit time is 5 days, the length of patrol on station required will be 35 days and the total length of patrol for each submarine will be 45 days (see Figure 11.1). The figures will vary with transit time as previously discussed. A shorter patrol might be achieved by the expense of faster transit speed; otherwise a shorter patrol time can only be achieved by moving to a three boat system (Figure 11.2). Then the cost of maintaining a single boat on station would be increased by the additional unit cost and running cost of an extra boat in the squadron.

A number of studies are usually required to arrive at an economic cost of such an operational scenario.

Maximum speed capability

11.5 The cost of speed has already been discussed and though high speed is always desirable some assessment of its value is warranted during concept studies. A possible assessment scenario is as follows. The submarine design under consideration may be assumed to be carrying out a surveillance role with a requirement to intercept targets crossing its barrier base line. (Figure 11.3). Based on the detection equipment to be fitted and the assessed characteristics of the target, there will be a range at which probable detection will be made, R, say. If the track of the target is at an angle θ to the bearing angle of detection there is an intercept problem to be solved. The shortest distance to target track will be $R \sin \theta$ and this is the distance the intercepting submarine has to cover whilst the target submarine makes an advance of $R \cos \theta$. For this simple intercept problem $\tan\theta = U_{sub}/U_{target}$. However, the submarine could increase the time to intercept by aiming off to an angle α on the diagram, which would increase the distance that has to be travelled by the target with only a small increase in the distance travelled by the intercepting subma-

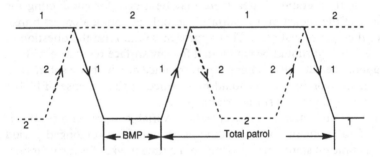

Fig. 11.1 2 Boat systems

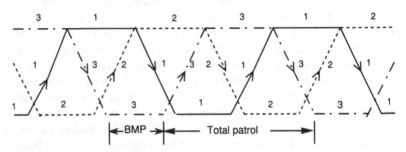

Fig. 11.2 3 Boat system

rine so that the speed required would be less. This is important for a conventionally powered submarine as the time endurance of the battery is considerably increased by reduction in speed. Thus if the submarine attempts to intercept at 20 Kts it may only be able to sustain this for half an hour, restricting the intercept distance to 10 miles whereas at, say, 15 Kts the endurance may be 2·5 to 3 hours, allowing an intercept distance of 40 miles. It should be remembered that the submarine may be required to carry out an attack at intercept so that the battery must not be exhausted during the closing run. Studies of this type show that for a conventional submarine the angle of target track is about 30° for successful intercepts of high speed targets. This angle, together with the probability of detection range, provides an assessment of the width across target track that the submarine can successfully prosecute. In turn, that consideration can lead to an assessment of how many submarines are required on station to provide a barrier force.

It is highly important that a continuous dialogue should occur between operational staff and the designers during the concept study stage related to the equipment and characteristics aspects, so that emerging concepts conform with the expectations of the operators and they are aware of the cost implications.

Options

11.6 Having established the primary requirements of the design, the next group of decisions may be encompassed by the term 'Technology Options', which represent the area in which the designer may most influence the concept design. Thus he may explore different materials for the pressure hull, alternative propulsion and energy storage systems and control arrangements. He may also explore the effect of more intensive

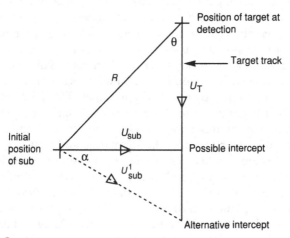

Fig. 11.3 Intercept geometry

automation with reductions in crew and alternative methods of noise attenuation. Such explorations may open up avenues for the introduction of innovation, but the designer has to keep in mind that the greater the degree of innovation the less reliable will be his data base and the more extensive will be the requirement for R & D before the design could be brought to fruition with minimum risk. That aspect may encourage investigation of alternative concepts, some of which may be regarded as 'safe' and others where the degree of uncertainty is high in the current state of knowledge. Although the riskier designs may not be acceptable if orders to build have to be placed before long, they may provide guidance to the R & D programme if they offer substantial benefits for the future.

Configuration

11.7   The designer may also choose to explore geometric configurations which depart from the 'norm'. For example, there may be reasons to consider a greater length/diameter ratio to accommodate a sensor base line or to improve surface performance. He may wish to consider the totally wraparound double hull or even twin or triple hull possibilities. He may wish to depart from 'cigar' form hulls with oblate or wing shaped forms to alter the manoeuvring characteristics. As discussed above such major departures could entail high risk and need extensive R & D before they could be realised but that should not inhibit concept exploration.

Standards

11.8   Another group of variants that can be explored are those which may be termed 'Standards'. Any design will have to meet a large number of standards which are usually specified in Design Manuals and are the result of investigations, some of long standing, which have arrived at acceptable levels. They cover issues like maintenance, reliability, quality assurance, fire, shock, noise hazards, factors of safety, levels of accommodation, weapon safety to mention just a few primary items. It may seem that these standards are set in tablets of stone and the designer has little freedom of choice to vary their requirements. Whilst that is generally the case, standards are continuously under review and in the context of a specific design may be open to exploration. Thus, for example, the method of arriving at the watch keeping bill and hence crew numbers, or the standard of accommodation on board, may have a significant effect on the size and performance of the submarine and so it may be worth testing out its applicability. There may be a case for reviewing overall factors of safety on diving depth if the control and speed of the design warrant debate about which safety standards could be relaxed in particular circumstances. One area where the designer may have a greater freedom to manipulate standards is the question of building and through life margins, predominantly for weight but also for machinery power.

It should be appreciated that the designer may be entering into fierce argument by suggesting changes to established standards, but it is incumbent on him to be aware of the cost of meeting set standards, bearing in mind that they might turn out to be unnecessarily high for the particular duty of the submarine whose design he is addressing.

METHODS OF CONCEPT DESIGN

11.9   Before any analysis of performance can proceed, an initial size and configuration of the submarine design must be generated as a start point. Although that could be a totally intuitive leap in the dark, for most designs some methodical approach is desirable and in fact necessary if designs of various options are to be properly compared. A number of ways exist at present and the increasing power of computers makes other ways possible. It is our purpose in this section to discuss some of the methods. Whilst we discuss them as if they were entirely independent, it is likely that a designer would utilise more than one technique to arrive at a viable concept.

Type ship method

11.10   The most widely known and well established method of Naval Architecture is that termed the 'Type Ship' approach. With this approach, the designer selects an existing, built design which most nearly matches the operational requirements of role, equipment and characteristics. In essence everything should be known about the type ship including any deficiencies that have come to light in its operation.

However, for the design of submarines the type ship approach is of limited value because of the constraints and limitations which the hydrostatic conditions of weight and buoyancy balance impose. If a new weapon and sensor outfit is required, the new outfit must fit within the overall configuration of the type ship and occupy no more volume within the pressure hull than the previous equipment. It should also require no more manning or electrical, hydraulic, air or water supplies than exist in the type ship. Those constraints amount to a change representing a modernisation of the existing class of type ship.

11.11   If the changes called for external form modification, then the propulsive performance of the vessel would be altered and possibly the control characteristics. These might be acceptable, with some degradation of vehicle performance, retaining the existing propulsion and energy storage system. However, it could be necessary to re-design the propulsor to match a new thrust requirement to the rotation speed and torque of the existing plant. If the degradation in performance was not acceptable then changes would have to be made in the propulsion plant which may result in additional volume and weight demands within the pressure hull.

If the above situation arose, or the new payload made greater demands on pressure hull volume or buoyancy, then the geometric configuration of the pressure hull may have to be changed. This would entail a whole sequence of modifications to recover the weight/buoyancy balance submerged and the main ballast tankage to obtain a correct surface condition. Changing the hull in that way would in turn bring about revisions to most of the operating systems of the boat (hydraulics, air, water, ventilation and electrical power distribution).

Increases in weight, without volume, may lead to a decision to use alternative higher strength material for the hull. This would probably result in a change in the safety margins for stress failure and buckling failure of the pressure hull and require a complete revision of plating and framing scantlings and spacing. It could also require a change in main hull bulkhead spacings which in turn caused revisions in deck layouts.

11.12   One form of modified type ship design that has been used with some success is the insertion of an additional module into the pressure hull. Ideally this module should be effectively self contained and neutrally balanced in order to avoid further changes to the original configuration. Such a change was successfully achieved in the World War II T Class of the Royal Navy, which was extended to add a further battery section and so improve its submerged performance. More recently the same approach has been adopted to introduce a Stirling Cycle AIP system into the Swedish Näcken Class.

From this brief exposition of the logic of change in a submarine it can be seen that it is very easy for simple modifications to a type ship to lead to major revisions of the design. The end result may be far removed from the original design and hence the risk factor, assumed to be contained by use of the type ship approach, may in fact turn out to be higher. Trying to avoid too many changes would probably lead to performance which was less than might otherwise have been achieved.

Hence it is the view of the authors that the type ship approach is of little value in submarine design and that it is generally better to start design *ab initio*. That is not to say that the useful data available from existing vessels should be ignored, rather that it should be used for guidance but not as a form of constraint in the new design.

The type ship approach may be applied in a more limited sense in which it is used just to provide a first estimate of size, from which the designer progresses with a clean sheet as to configuration.

Building block method

11.13   A less constrained approach, which still draws heavily on existing practice and data, is what might be called the 'Building Block' approach.

In this approach the boat is 'assembled' from component 'blocks' for which weight and space data are known. Thus for the power plant an existing electric propulsion motor with its ancillaries may be selected; similarly a diesel generator plant (say twin diesels) with its associated equipment may be chosen; also an established weapon discharge system and suitable Command and Control compartments may be selected.

From this assembled data a basic volume requirement can be assessed which, with calculated tankage, will provide a pressure hull volume and a knowledge of the internal weights. The hull scantlings can then be determined and the external MBTs and free flood spaces estimated to give the required surface condition.

In that way the size and configuration of a possible option can be derived. The next step is to carry out an analysis of the vehicle performance of the postulated design and to compare the results with the operational requirements. If a satisfactory comparison is not achieved then alternative choices of equipment can be progressively made until an acceptable performance is reached. The process is essentially iterative and though developed in pre-computer days it is very suited to modern computational methods provided a satisfactory data base is available. With the computer the iterations can be achieved rapidly and the role of the designer is to make experienced selections. Though the method is less constrained than the type ship approach, it is nevertheless very dependent on the quality and variety of the data base and that limits the innovation which could be introduced into the Concept.

11.14 In order to make initial selections of equipment some preconceived idea is required of the probable size of the vessel. Though subsequent iterations may revise the earlier estimate there are indications that a process is at work which tends to be self-satisfying, i.e., that the original expectation of size leads to a vessel of about that size. The algebraic analysis offered later in this chapter gives some clue as to why there is this self-satisfying behaviour. Another explanation is that the selected 'building blocks' are less interactive with the other elements than they should be. In support of that contention, we have found that the process can be less sensitive to changes than might be expected. Despite those drawbacks, the building block method remains a useful way of generating designs to meet customer's requirements based on current technology. It can be extended to provide more forward-looking designs by the creation of 'pseudo' data base building blocks. Thus, by carrying out a partial detailed design of, say, an AIP plant, the information generated can be added to the data base. However, if a computer aided design system is being utilised, particular care has to be exercised in the management of the database files to avoid confusion between real and created data. Another way in which the data base can be extended is by including data created by conducting design studies using the system. Again, care must be exercised in the use of purely 'paper' designs as a source of data for further designs, as the degree of validation is much less than data derived from 'built' designs. Even with this extension to the data base, the building blocks are discrete units and if a change is required the designer must select the next or nearest unit available in the data bank. For some units this is realistic, for example, diesel engines are usually only available in particular units, and consoles are normally available in specific sizes. However, many systems can be tailored to suit the demand and therefore have a more continuous relationship between requirement and size.

A further stage of manipulation of a data base is, therefore, to develop relational equations to estimate the weight, volume and power demands of the system in response to input requirements. These may be generated by formulating logical algebraic algorithms connecting design parameters. The existing data base may then be used to determine the coefficients in the relational equations. Regression analysis is frequently adopted to perform this function.

Graphics method

11.15   Some designers contend that the broad manipulation of volumes to arrive at size is unnecessary and that it is the achievement of a satisfactory, workable geometric arrangement particularly at the fore-end which determines the ultimate size of the vessel. Therefore their start point is the development of arrangement drawings, as discussed in Chapter 7. The authors would agree that the establishment of a satisfactory geometric arrangement must come early in the concept design phase if the design is to be considered viable. The arithmetical manipulation of volumetric demands is no guarantee that a design can be achieved. However, they believe that even an arrangement drawing must start from some, albeit tentative, framework of space and this requires an initial estimate of total volume and associated length and diameter (though dimensions may have to be modified subsequently to fit the specific measurements of equipment).

To some extent advocacy of a direct leap to geometric sizing arises from experience in design of vessels where a major proportion of the submarine is fixed. Where there has been a lengthy development programme for the main propulsion plant this cannot readily be changed and hence half the submarine is virtually fixed in volume and dimensions and the problem is one of defining the other half. To a large extent the performance of the vessel is fixed by the plant available and the design is limited to keeping fore-end size down and achieving the best hydrodynamics in the circumstances. Where there is greater choice in the separate equipments of the power plant it can be more worthwhile to start with manipulation of gross volumes in order initially to size and dimension the design.

11.16   As well as using the computer to conduct fast arithmetical calculations, its value can also be utilised for exploring configurations and layouts using the powerful graphics capability of most modern computers.

At a 'simple' level the computers can be used as a draughting tool. Using a menu tablet the external configuration and pressure hull can be drawn, which generates a data base and so enables deck and compartment shapes to be called up and enlarged to conduct more detailed layout studies. The geometric coordinates allow tank volumes to be computed and centres of gravity of main items to be identified and accumulated as steps toward the total balance calculation. Standard items of equipment can be pre-drawn and scaled or moved about so that many possible layouts can be explored. This provides a considerable speeding up of previous hand draughting methods. With powerful computer packages, access routes and the movement of stores or replacement parts can be examined. It is also possible to generate internal perspective views of compartments which provide the user with a clearer picture of the final layout.

Whilst such aids are available they should not be allowed to obscure the basic underlying need for thoughtful design and should only be used where they really do serve to facilitate the design process. Thus the manipulative draughting tool is valuable to the early evaluation of possible configurations and layouts.

11.17   In the context of concept design there is a need to link the numerical computation of size, weight and balance with the geometric evaluation of configuration. A way in which this could be done is by the automation of the previously described 'Flounder' diagram in Chapter 7. Once the computer has achieved a space/weight balance it becomes possible to generate the boundaries of such a diagram. These can be generated both for the internal volume of the pressure hull (shown as an area in the diagram) and of the external volumes. From the input values of space requirements and the computed values of tankage, propulsion plant, batteries, etc., the area within the diagram can be manipulated to dispose these spaces within the diagram and so assess the residue of space (positive or negative) that might arise due to the juxtaposition of spaces. A residue may arise because, though the total volume required has been determined, it may not be distributed in a manner which allows achievement of a suitable spatial layout of compartments and tanks. For example, since the weapons compartment has to be all together in one part of the hull, to reach a situation in which 80% is in one place and 20% elsewhere is unlikely to be acceptable.

CONCEPT SIZING

11.18   Another approach to initial concept sizing is based on the computer and its ability to iterate rapidly. A form of it is used at University College London for teaching students the basics of submarine design and is designated SSCON, which in many ways resembles current CAD packages used for 'real' designs. SSCON utilises the space/density approach discussed in Chapter 4. It was originally developed because of the non-availability of 'real' data bases as described in earlier paragraphs. However, in some ways this has made it quite versatile because it is more continuous in its manipulation of space and weight and so allows some scope for innovation in design as it is not tied to discrete blocks or packages.

Over a number of years a series of dependency algorithms have been incorporated which generate 'realistic' demands for volume, weight and power based on the input parameters, and an estimate of cost based on the assessed group weights.

The essential inputs to estimate the size of the concept are the demands in weight and space of the payload. The payload is defined as the equipments required to perform the operational role of the submarine, i.e. the sonars, communications, electronic warfare and radar sensors and associated processing equipment; weapons, with their fire control system, discharge system and reload storage; the command and control system to link the separate systems into a total weapon system. In addition to the payload demand for volume and weight there will be associated requirements for electrical power, hydraulic and air supplies and manning for the operational posts and maintenance. There will also be water compensation tankage for the discharged weapons. The total of these requirements constitutes a pre-determined payload which is independent of the size and configuration of the submarine to carry them.

The rest of the components which make up the totality of the submarine are, either directly or indirectly, dependent on the size of the boat and its performance characteristics. There is no difficulty in feeding the performance requirements into the calculation, but the size of the boat, which is the ultimate objective of the computer computation, has to be known in order to proceed. It is therefore necessary to have a preliminary estimate of size.

11.19   For a conventionally configured design, one approach to providing this initial estimate is to use a payload/hull volume fraction. As was seen from the tables in Chapter 4, the volume occupied by the payload is approximately 30% of the volume inside the pressure hull. If a volume controlled design is envisaged it is therefore possible to make a preliminary estimate of the internal volume of the pressure hull by the simple algorithm:

$$\text{Pressure Hull Volume} = \frac{\text{Payload Volume}}{0 \cdot 3}$$

Although it would be possible to make a weight controlled design approach by using a payload/weight fraction, as discussed more fully elsewhere, the authors recommend an initial volume controlled design approach even if it subsequently turned out that the sizing of the vessel was weight controlled.

Given this estimate of required usable volume inside the pressure hull one can proceed to obtain estimates of the further additions of space within the pressure hull.

11.20 To the 'dry' volume must be added a figure for internal tankage. This will consist of domestic tankage, which in the first stage of calculation can be taken as proportional to the Pressure Hull Volume, PH Vol..

There will also be internal tankage for lubricating oil and ready use fuel. Though these are not dry spaces they are normally included in the Machinery Volume fraction and covered by the overall Payload Volume fraction.

An important additional volume within the Pressure Hull is the space required for the Trim and Compensating (T & C) Tanks. The determination of the T & C system tank sizes and location is a separate design exercise as discussed in Chapter 3 and described in detail in Appendix 3. For the concept sizing exercise it can be simplified. It will be seen from the Appendix that the extremes of weight variation are dictated primarily by the range of density and, in the extreme light condition, the consumption of usable stores. Therefore, for computational purposes an algorithm of the following form may be used:

$$\text{T\&C Vol} = \left[\frac{\text{PH Vol} (\rho_{max} - \rho_{min})}{\rho_{sw}} + \frac{W_T \text{ of Stores}}{\rho_{sw}} \right] \frac{1}{(\text{Utility factor})}$$

where $(\rho_{max} - \rho_{min})$ is the required range of sea-water density.

The weight of stores can be estimated from the crew size and patrol endurance. Whilst only the crew required for the payload has so far been established a total manning estimate can be made from the initial size estimate; alternatively a maximum crew size may have been decided as part of the requirements. A typical algorithm for stores weight takes the form:

$$\text{Stores } W_T \propto (\text{No. of Crew}) \times (\text{Patrol Endurance (days)})$$

The utility factor is a fractional value to allow for internal tank structure and 'ullage', approximately 0·95 each, i.e the margins full and empty as some residual water or air space will be left.

For simplicity one could take Vol of T & C = 0·045 Surfaced Displacement Volume/Utility Factor.

11.21   The addition of these calculated tankage volumes provides a total internal Pressure Hull Volume (PH Vol_{int}) requirement. To determine the external Pressure Hull Volume (PH Vol_{ext}), the calculated requirement must be increased to allow for the shell plating and framing with only partial utilisation of the space between frames. Typically the External Pressure Hull volume will be:

$$PH\,Vol_{ext} = PH\,Vol_{int} \times 1·15$$

This figure effectively determines the 'submerged' displacement of the submarine though, as discussed in Chapter 3, the submerged displacement normally is defined as including MBT water. In the usual terminology, the buoyancy of the PH Vol_{ext} obtained as above is close to the surfaced displacement of the submarine.

It is now possible to determine a volume of the external MBT by the use of the Reserve of Buoyancy fraction. This is to some extent an optional choice of the designer but will usually be between 10–15% of the PH Vol.

Therefore one can calculate

$$MBT\,Vol = (PH\,Vol \times ROB)/Utility\,factor$$

The Utility Factor is again included to allow for internal MBT structure and an amount of residual water left in the bottom of the tanks when fully blown. It is prudent to increase this factor to allow for high pressure air bottles stored in the tanks and possibly for permanent ballast if this is to be stowed externally.

Adding the derived figure for MBTs produces a volume whose buoyancy is normally termed the submerged displacement of the submarine. Yet another volume is required for subsequent calculations and this is the Form volume, i.e., the total volume within the overall submarine envelope. At this stage it is not certain how much additional free flood volume will be necessary in the final configuration but allowance must be made for external volume at the bow and stern plus the Bridge Fin and the Casing if the latter is envisaged as a separate extension along the top of the hull.

For calculation purposes the free flood volume may be allowed for by adding, say, a further 15% to the previous volumes so that:

$$Form\,volume = (PH\,Vol + MBT\,Vol) \times 1.15$$

With these three volumes and the vehicle performance characteristics it is now possible to proceed with more detailed analysis of the volumes and weights of the submarine vehicle equipments which are dependent on these factors. (It may be noticed from Chapter 3 that the variations in buoyancy due to density changes should strictly be applied to the Form volume which has been derived subsequent to the T & C Tank Volume. It is, however, usual practice to size T & C systems on the PH Vol rather than on the Form volume. If it were desired to correct for the discrepancy, the figure for PH Vol used in the algorithm for T & C Volume could be enhanced by the subsequent percentages for MBT and Form volume.)

The main components now to be estimated are those of the propulsion and energy storage systems which until now have been assumed to occupy the estimated space within the pressure hull.

Assuming a standard diesel electric propulsion system, the algorithms described below may be used to estimate the volumes and weights of the major components.

Propulsion motor

11.22   The size of the motor is governed by the maximum power required to be transmitted to the propulsor. This will normally be dictated by the maximum submerged speed, U_{max}. The effective power, i.e. the thrust power for a given speed, may be estimated by the algorithm:

$$\text{Power}_{\text{eff}} = K_p \times (\text{Form Vol})^{.64} \times U_{max}^{2.9}$$

For metric units and a typical form for submerged performance, the factor K_p takes a value of about 20. If the form to be adopted departs significantly from the 'ideal' discussed in Chapter 6 then this factor may be increased by a small amount, i.e. it may be modified appropriately for the L/D ratio of the form.

The power required to be transmitted by the propulsion motor will be greater than the thrust power by the hull efficiency η_H, the propulsor efficiency η_0 and the transmission efficiency η_s, so that:

$$\text{Motor Power} = \frac{\text{Power}_{\text{eff}}}{\eta_0 \times \eta_H \times \eta_s}$$

At this preliminary stage little is likely to be known of the details of the propulsor or stern configuration. For a single axial propulsor $\eta_0 \times \eta_H$ may be assumed to be about 0·75 and $\eta_s = 0·98$. For a twin screw vessel the power will be divided into two shaft powers and $\eta_0 \times \eta_H$ will be in the vicinity of 0·6.

The weight and volume of the propulsion motor and its ancillaries may then be estimated from the required power and the shaft revolutions, i.e. to meet the effective torque requirement.

Battery sizing

11.23   The next figure to be estimated is the weight/volume required for batteries. The sizing of the batteries depends upon the energy drain during a single submergence of maximum duration. This energy drain will be due to two components, namely, the energy demand of propulsion at submerged endurance speed and the energy demand from within the submarine, i.e. the Hotel Load.

For a single submerged endurance speed the power can be estimated from the previous formula, substituting the submerged endurance speed U_{sub} for U_{max}. The total energy required will be this power estimate multiplied by the time in hours of the submergence.

If a pattern of speeds is required whilst submerged then the total energy requirement will be $\Sigma P_{ui} \times t_i$ where t_i is the time in hours at speed U_{subi}.

The hotel load will be the power required for the payload in normal patrol state plus the power required to run auxiliary machinery and the ventilation and air conditioning load in normal patrol state. The payload demand can be derived from the input information on the required payload, but at this stage no detailed information is likely to be available for the submarine systems. For estimating purposes this figure can be assumed to be proportional to the PH Vol.

Thus hotel load may be estimated from the following algorithm:

$$\text{Sub HL} = 0.75 \text{ HLPAY} + 0.075 \times \text{PH Vol}$$

The total battery energy requirement can then be estimated and turned into a calculation of the Number of batteries:

$$\text{No. of batteries } (N_{Batt}) = \frac{\text{Submerged energy}}{\text{Stored energy/ cell}} \times (1 + x)$$

where $(1 + x)$ is a battery safety margin.

From the forgoing estimates it is now possible to estimate the maximum time that full power could be sustained assuming a fully charged battery:

$$\text{Time at max speed} = \frac{N_{Batt}}{P_{Max}} \times K_{MB}$$

The factor K_{MB} makes allowance for the reduction in energy available if the batteries are discharged at maximum rate.

Diesel generators

11.24   The diesel generator plant sizing is based on the power which is required to be delivered by the diesel engines whilst snorting. In this condition the diesel generators have to provide power for three distinct purposes. These are:

Propulsion power at snort speed and depth
Charging the batteries
Hotel load in the snorting condition

The first of these can be estimated as before but using U_{snort} in place of U_{max}; the coefficient may also be increased to allow for some wavemaking resistance at snort depth.

For the second purpose, assuming standard batteries, there is a maximum current limitation on charging, so that this power is fixed by maximum current (I_{max}) and the number of cells already determined.

The third purpose is the hotel load already discussed, but in the snorting condition it is to be expected that this load will be reduced as the boat will be ventilated from atmosphere during snorting. Hence the estimation of required diesel power takes the form:

$$P_{Dies} \left[(1 + m)\,(I_{max} \times 2 \cdot 4 \times N_{Batt}) + \frac{P_{snort}}{\eta_{el}} + HL_{snort} \right] \frac{1}{\eta_{Conv}}$$

where m is introduced as an oversizing factor to allow a margin and efficiencies η_{Conv} and η_{el} are introduced to allow for the diesel generator conversion of power to electrical demands.

At this stage it is possible to make a selection of suitable diesel generator plants that would provide P_{Dies} and to use the data base to obtain the volume and weight requirements. Alternatively, for initial sizing, an algorithm of a more continuous nature can be used to avoid specific selection of plant. That approach allows for iteration until a balanced solution is obtained at which stage a specific selection can be made. It is found from existing data that weight and space requirements do not increase linearly with power, and so in the algorithm it is suggested that the plant requirements are taken to vary with P_{Dies} to the power of 0.7.

11.25   Also at this stage an estimate can be made of the important operational factor of Indiscretion Ratio (IR). This is defined as the time snorting divided by time of transit:

$$IR = \frac{T_{Snort}}{T_{Snort} + T_{sub}}$$

It can be seen that the time of snorting is fixed by the time taken to charge the batteries at the allowable charging rate, i.e.

$$IR_{min} = \frac{T_{chmin}}{T_{chmin} + T_{sub}}$$

It follows that the indiscretion ratio is independent of the selected diesel generating plant, provided it is large enough to supply the maximum charging current. The IR can only be reduced either by increasing battery size or by reducing the submerged power requirement, which is generally achieved by reducing the propulsive power. The latter can be influenced by designing a low resistance hull but is primarily dependent on having a low speed of submerged operation.

The foregoing account is related to use of a standard diesel electric power plant. It emphasises the value of an efficient Air Independent System to provide power submerged because it reduces the need for snorting. Some forms of high energy density battery do not have a charging current limitation, which makes it possible to increase the diesel generator power and reduce snorting time.

Fuel requirements

11.26   Though part of a patrol will be conducted submerged operating on batteries, the energy required will have been provided while the diesel engines were running. Therefore the fuel consumption will be dictated by the running time of the diesels and their power setting. Consequently, the fuel stowage requirement can be determined from the summation of the diesel running times expected during a patrol and the specific fuel consumption of the engines at the power setting. Hence an algorithm for the computation of fuel capacity would take the form:

$$\text{Fuel Weight} = P_{\text{Dies}} \times \text{Time} \times \text{SFC}, \text{ where SFC is in kg/kw}$$

However, in the preliminary estimates of size some of this information may not be known. A simpler first estimate can be obtained from the requirement of total range and the power for a mean speed of advance.

$$\text{Hence Fuel Weight} = \left(\frac{\text{Range}}{\text{Mean Speed}} \times P_{\text{mean}} \right) \times K_{\text{fuel}}, \text{ where } K_{\text{fuel}} \text{ is}$$

an average conversion efficiency of fuel energy into power.

CONCEPT SIZING WITH AIP SYSTEMS

11.27   The concept sizing process that has hitherto been described is based on a conventional diesel electric propulsion and energy storage system. Some differences arise if alternative systems are envisaged in the concept.

Nuclear propulsion

This alternative cannot be addressed in detail but it will be apparent that for the most part a nuclear submarine design revolves about what amounts to a fixed power plant and propulsion system; the long development and testing time of that installation makes it virtually invariant in the design process. Submerged endurance is no longer a significant energy storage problem, though it remains one as regards the provision of accommodation, ventilation, air conditioning and stores for the crew for the exceptionally long periods of sustained submergence attainable. The propulsion plant takes up a large proportion of the hull and virtually fixes the size of the submarine within relatively small variations. Performance assessment is primarily one of establishing what the plant will achieve.

For future design concepts there may be more flexibility to vary the turbines and shaft power transmission systems so as to tailor them to postulated design requirements. Such studies are more likely to be used to guide future plant developments rather than for the purposes of design for building on a short time scale.

There have been proposals for smaller nuclear plants producing power in terms of kilowatts rather then megawatts. Such plants can reasonably be regarded in the same context as other so-called Air Independent Systems.

AIP systems

11.28   Apart from the small reactor just mentioned the current possibilities for AIP are:

   (a) Recycle diesel

   (b) Stirling cycle

   (c) Fuel cell

   (d) High energy density (HED) batteries

The last of these is not strictly an AIP system but does bring about changes in the design process. Research findings indicate that on a comparative volume basis the high temperature battery, for example, can store 3–4 times the energy of a conventional lead/acid battery. That capability implies that if a submarine were designed with the same volume of batteries it could either increase the submerged endurance by a similar factor for the same patrol profile, or allow higher speeds. As has been already observed there is less restriction on charging rate with the high temperature battery so that the time to recharge is not commensurately extended, and considerably better indiscretion ratios can be achieved.

It may be that operational requirements do not give rise to the need for realisation of this extended performance, in which case a smaller HED battery volume would be required than for the lead/acid type. That could give rise to problems in submarine layout because the space saved by the battery in the lower level of the boat is not readily usable for other purposes. Using the concept sizings programme with data for HED batteries may result in a PH Vol which is inconsistent with the architectural layout of the boat. Some of the space made available could be used for additional stores, though these are more dictated by the duration of the patrol than by submergence time. One possible solution is to include more internal fuel tankage where previously fuel was stowed outboard. No doubt some form of arrangement could be found, differing from that currently adopted. This apparently simple change illustrates yet again the interactive nature of the design problem.

The other systems pose different problems. They all require fuel, sometimes of a special type, e.g. methanol, but this can be included in the general assessment of fuel for the total patrol. The significant difference is that they also require the carriage of an oxidant. Therefore in the sizing programmes an additional volume and weight allowance has to be made for oxidant. If it is carried in the form of liquid oxygen (LOX) then about twice the volume of LOX has to be allowed to cater for the space for the thermal insulation. In addition compensating tank volume must be provided to maintain trim as the LOX is consumed.

The recycle diesel system will use the main engines but additional space is required for the recirculated gas, mixing and discharge of combustion products. (The latter requirement applies whatever type of AIP system is employed)

11.29   The fuel cell or Stirling cycle are virtually additional systems to those usually fitted in a conventional diesel electric submarine and therefore additional space has to be allowed. The capacity of plants now becoming available is not sufficient to replace either the diesels or the battery. Their main function is to extend the submerged endurance of the submarine. At a moderate speed they can provide sufficient power for propulsion and hotel load so that the battery need not be discharged at all or, at enhanced speed, would have a smaller discharge rate. However, for full speed submerged the submarine is still dependent on battery power and voltage. Thus although the battery sizing may no longer be dictated by submerged endurance it may still be of the same size to provide the high speed performance.

Depending on the degree to which AIP plant provides submerged energy the diesels may only be required to charge batteries on an occasional basis. It may therefore be acceptable to reduce the size of diesels and extend the time taken to fully charge the battery should it be expended. However, that would be likely to reduce speed performance on the surface.

OTHER SIZINGS
11.30   With the main propulsion elements sized for weight and volume the designer may almost be in a position to reassess the preliminary sizing of the submarine design. However, a few more estimates can be made before taking that step.

The sizing of the propulsion system paves the way to assessing the manning of the vessel. Thus a re-estimate of crew size can be made in the form:

$$N_{\text{men}} = N_{\text{Payload}} + N_{\text{Prop}} + N_{\text{Hull}}$$

The first term was part of the initial input and numbers can be generated for the other terms. From the standards to be adopted for accommodation an estimate of what might be termed the domestic weight and volume can be made. The auxiliary systems can be assessed as a simple proportion of the hull volume, which is usually good enough for initial sizing as to a large degree the systems are dependent on the hull size.

At that point all the demands for weight and volume within the pressure hull will have been more closely estimated, together with the fuel requirements. The latter may be split between internal and external tankage depending on the configuration. However, one major weight item still remains to be assessed and that is the structural weight.

Pressure hull weight

11.31   As discussed in Chapter 5, the design of structure to withstand external pressure is a more complex problem than that with internal pressure. Nevertheless, the 'boiler formula' gives a good guide to the sizing of the pressure hull structure of a submarine. The boiler formula for a cylindrical shell is:

$$\sigma = \frac{p \times D}{2t}$$

where p = design pressure, D = hull diameter, t = shell thickness and σ = allowable stress.

Rearranging this formula gives:

$$t = \frac{p \times D}{2\sigma}$$

and so the shell weight is given by:

$$\pi Dt \times \rho_s \times d\ell \text{ per unit length} = \frac{\pi D^2 \times p}{2\sigma} \times d\ell \times \rho_s$$

where ρ_s is the density of steel or whatever other material is adopted.

However, the pressure hull volume/unit length = $\pi D^2 \times d\ell$, giving the pressure hull shell weight in the form:

$$\text{Weight} = \frac{p}{2\sigma} \times \text{PH Vol} \times \rho_s$$

To a reasonable degree of approximation for preliminary purposes, the pressure hull frames can be taken as proportional to the shell. It follows that pressure hull weight is independent of diameter as such and can be related to PH Vol, the design diving depth and the design stress to be adopted for the material. Thus an algorithm for the initial estimate of pressure hull structure weight can take the form:

$$\text{PH } W_T = K \times \frac{\text{design diving depth}}{\text{design yield stress}} \text{ PH Vol} \times \rho_s$$

Depending on the material properties of yield stress and modulus the relationship with design diving depth may not be linear.

The internal structure is not subject to diving depth but for initial estimating purposes may be also taken as proportional to PH Vol. Similarly external structure may be taken as proportional to the external volumes which have been estimated.

11.32   Having made direct assessments of the volumes required for the main functions of the submarine and estimates of the auxiliary function demands, it is now possible to total them up to find a revised requirement for PH Vol and hence for MBTs and Form vol. If the design requirements are fairly standard the new figures may be expected to be quite near the provisional estimate made using Payload fraction. However, if there were significant differences in requirements the new figures could diverge appreciably from the provisional estimate. This could occur if a larger payload requirement were demanded on a modest vehicle performance, or alternatively if a vehicle performance were required with enhanced maximum speed and/or submerged endurance. These differences would alter the payload fraction relationship, and it would be necessary to re-enter the calculations and re-iterate based on the new figures for the basic volumes. Several such iterations might be necessary before the estimation process converged to steady or near steady values. It is usual to accept that convergence has been sufficiently approached when the figure for PH Vol is within about 2% of the previous value. A flow diagram for an iterative sizing program is shown in Figure 11.4.

Balance check

11.33   At the same time as volume requirements are being assessed, weight information is also being generated, which allows a further check to be made on the state of the weight/buoyancy balance. The check is usually made by comparing the buoyancy of the pressure hull volume with the summation of component weights supported by this buoyancy. For a volume controlled design the buoyancy of the pressure hull should exceed the total weight to a sufficient extent to allow for an acceptable amount of permanent and margin ballast. This simple balance is likely to be somewhat pessimistic as it does not take account of the buoyancy provided by external structure and equipment and also ignores the buoyancy of external fuel displacing sea water in the standard stores condition. However, it affords a sufficient check to ensure that the submarine design when it converges has not moved into a weight governed condition.

If a weight/buoyancy imbalance does appear in the calculations then steps have to be taken to correct the design. A weight saving strategy may be adopted, usually by reducing the pressure hull weight, which may be accomplished either by reducing design diving depth or by changing to a higher strength material for the hull. These are the two major parameters in the hull weight algorithm, but it may be necessary to change the K factor to suit the new material. For example, a higher yield steel may need more stiffening to improve the buckling pressures commensurate with the higher stress loading.

Another alternative would be arbitrarily to increase the pressure hull volume over the calculated demand to provide more buoyancy. Ideally this would provide more comfortable spaces on board, e.g. improved accommodation standards. However, as we have shown, many of the algorithms for weight and space are related to pressure hull volume. Thus much of the increase would be eaten up by the consequent increased demands. It is very easy for the process to get out of hand and require a disproportionate increase in the size of vessel to achieve a modest improvement in weight/buoyancy balance.

11.34 This tendency for a design to 'grow' highlights the self-satisfying nature of submarine design discussed earlier. In a design which remains volume controlled the effect may not be so obvious but it exists nevertheless. It may seem that the crude preliminary sizing used to start the design cycle is not important, as it is replaced on second iteration by more detailed assessment. Nevertheless, because so many of the dependent algorithms relate to size, a self-satisfying process is in operation. In that way if a large initial sizing is adopted, the calculations tend to confirm a large solution. It is possible that a smaller convergent solution exists but has not been identified by the calculation process.

To avoid an over-large outcome from the calculation it is advisable to start with a small initial estimate and be forced larger only if the iteration process requires that to happen. The question is, how small should the provisional estimate be?

If the calculations are done manually it is desirable to make a good first estimate so that only a few iterations have to be carried out to achieve convergence. With the availability of fast computing that approach is unnecessary as many iterations can be conducted in a short time. To keep the design as small as possible a very small initial estimate should be used. It is suggested that this first estimate to 'kick start' the computation could be just the payload volume demand. Clearly a pressure hull volume which only provides space for the payload is far too small and makes no provision for the propulsion system. However, the first iteration will generate volume requirements for propulsion and batteries so that a somewhat larger pressure hull volume is generated for the start of the next iteration. The propulsive demands will then be increased and so a yet larger pressure hull volume will be generated for the next iteration. By rapid repetition the sizing will 'grow' to reach a stable value. There are grounds for confidence that this stable value will be the smallest solution to the requirements. The approach also provides greater sensitivity to changes in design.

Choices of parameters

11.35   It may be considered that some of the algorithms in the suggested concept sizing process are over-simplified and that, with the computer power available, more exact calculations could be conducted. For example, why not carry out a full structural analysis and determine the weight of structure in detail. If this were to be done more information would be needed on the overall configuration of the hull and decisions on these parameters would, in our view, predicate the concept prematurely. It is important to recognise that arriving at an estimate of size and cost at the concept stage is only a start to the design programme. Once a size and weight/buoyancy balance have been identified, it is then necessary to find out whether a length, diameter and hull configuration can be established complying with those figures which permit a satisfactory arrangement of spaces and equipments to be achieved. With that necessity go the further hydrostatic conditions of longitudinal balance, satisfactory transverse stability and surface flotation. There would seem little value in elaborating the concept sizing process whilst those factors remain to be resolved.

CHARACTER OF ITERATIONS

11.36   The separate dependency algorithms make it very difficult to see directly how the effects of changes influence the outcome. A simplified approach may help to illustrate the nature of the process which is taking place.

It is possible, using a simple algebraic analysis, to demonstrate how the interactions operate in the design of a vehicle for which the weight of its components is the determinant of its size. Although the submarine case is more complex because space requirements also or mainly influence size, we consider the analysis that follows is instructive.

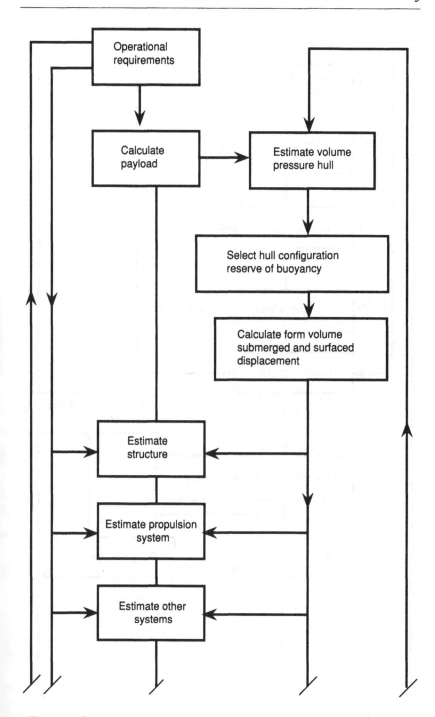

Fig. 11.4 Submarine sizing procedure

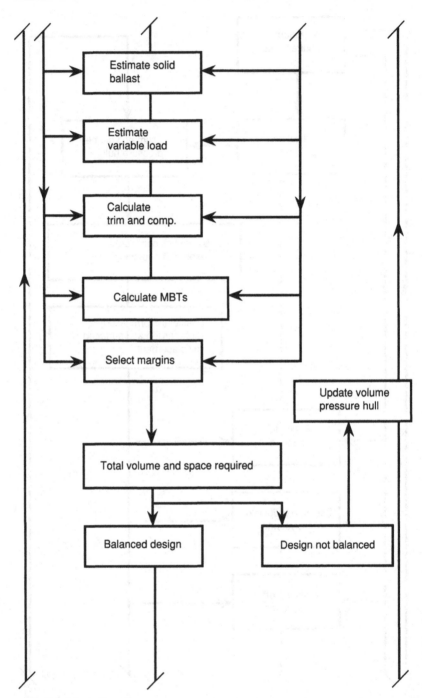

Fig. 11.4 (continued)

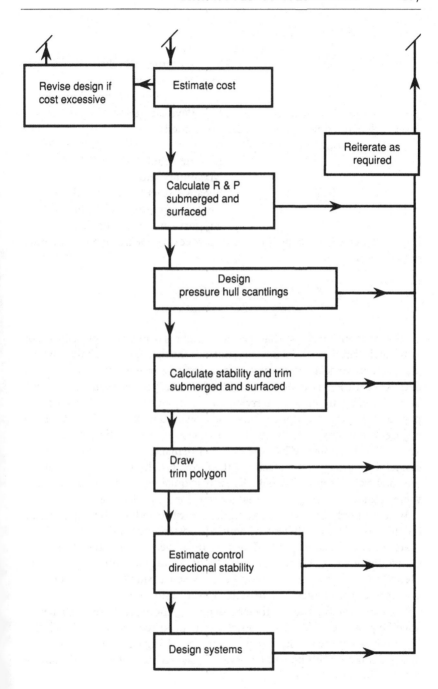

Fig. 11.4 (continued)

We start with the relationship:

$T = S + M + P$ where T = Total weight
S = Structural weight
M = Machinery weight
(including electrical power)
P = Payload weight

A set of dependency expressions can be formed:

$S = \alpha_T T + \alpha_M M + \alpha_P P$ because Structure depends upon total size plus provision to support Machinery and Payload

$M = \beta_T T + \beta_P P$ because Machinery depends on total size for propulsion plus support demands of Payload.

If the coefficients are constant, i.e. the process is linear, then by algebraic substitution:

$$T = \frac{P(1 + \alpha_P + \beta_P + \alpha_M \beta_P)}{1 - \alpha_T - \beta_T - \alpha_M \beta_T}$$

This relationship shows that the total weight is dependent on payload not only the declared weight of payload, but also the demands which the payload places on the vessel for structural support and power requirements. The sum of those coefficients can often be greater than one, yet it is frequently ignored in considering the payload carrying capability of the vessel. Indeed some payload equipment designers keep their own equipment weight down by relying on the submarine designer to provide a large proportion of the necessary payload support. Whatever the total demand of the payload, it is all divided by an exaggeration factor of the form $(1 - A)$, where A is the sum of the total size dependency coefficients. It is quite easy for A to approach a value of unity, resulting in a very large vessel for the payload carried.

It accordingly behoves the designer to try to reduce the dependency coefficients. The coefficient α_T depends on strength and stiffness of the material and the structural efficiency. Improvement of either to reduce α_T would lead to higher complexity of production and higher cost. Hence there is a possible trade off between a smaller vehicle at high cost/tonne and a larger vehicle with a lower cost/tonne.

Similarly for β_T, the coefficient depends upon propulsive efficiency, specific power density of machinery and speed requirements. While for the first two there is likely to be a degree of higher cost involved, the speed demand (with its cubic relation to power) can be a very dominant element in the solution.

11.37   A slight elaboration of the foregoing equations illustrates the fine complexity of the design situation.

If a further breakdown of the total weight is made in order to include manning then:

$$T = S + M + C + P$$ where C represents manning weight including stores and effects

where:

$$S = \alpha_T T + \alpha_M M + \alpha_P P$$ assuming C does not make significant demands on structure (though in fact there will be tanks and minor structural compartmentation associated with C)

$$M = \beta_T T + \beta_C C + \beta_P P$$ to reflect that the crew place demands on power supplies for heat, light, air conditioning and refrigeration

$$C = \gamma_M M + \gamma_P P + \gamma_T T$$ to take account of the fact that manning is required for machinery and payload and may also depend on size of vehicle.

When these are solved algebraically:

$$T = \frac{P\left[(1 + \alpha_P + \dfrac{\alpha_M(\beta_C\gamma_P + \beta_P) + (\beta_P + \beta_C\gamma_P) + (\gamma_M\beta_P + \gamma_P)}{(1 - \beta_C\gamma_M)}\right]}{1 - \alpha_T - \left[\dfrac{\alpha_M(\beta_C\gamma_T + \beta_T) + (\beta_T + \beta_C\gamma_T) + (\gamma_T + \gamma_M\beta_T)}{1 - \beta_C\gamma_M}\right]}$$

This is a far more complex relationship but is of similar form to the previous simplified dependency equations. Now, however, there are added terms both top and bottom, and internal exaggeration factors have appeared because of internal cross dependencies. Thus on the top line there is the term $(\gamma_M\beta_P + \gamma_P)$ which is the direct manning of the payload plus the manning for power supplies demanded by the payload. The whole of the manning requirement is exaggerated by $(1 - \beta_C\gamma_M)$ i.e. the manning of machinery to support the crew.

11.38 We do not suggest that this analysis is itself a means of solving the design problem. It is virtually impossible to identify these algebraic components in a real vessel. Furthermore, the coefficients are not constant and dependencies are not linear and may have step functions, e.g. due to choice of available power plant units. Nevertheless, it serves to illustrate the complexity of the interactions in a submarine. Attempts to elaborate further with more detailed system breakdowns and dependencies lead to enormous algebraic expressions which are difficult to handle. However, the same format of:

$$T = \frac{P(1 + (\text{Payload demands}))}{(1 - \text{Vehicle demands})}$$

emerges with many internal exaggeration factors.

The foregoing algebraic analysis shows some important features. If the vehicle performance characteristics are demanding, e.g. high speed or deep diving, the denominator of the generalised equation will approach zero. Hence a design could become very large whilst the payload fraction becomes insignificant. Effectively the vehicle would be propelling its propulsion plant. The previous approach assuming a 'normal' payload fraction could in those circumstances be in considerable error. Alternatively a design may be required with a very modest vehicular demand and with the payload representing a more dominant fraction of the total. Figure 11.5 illustrates those characteristic changes. The described concept sizing programme does not work too well with such radical departures from the 'norm'; although it might be expected to converge whatever the initial estimate it does not react fully to the changed input. It appears that the initial estimate dictates the size region within which the iteration finds a balance. This is particularly true where there are steps in the selection of specific machinery units.

Beyond the concept

11.39   By processes such as those described, a series of concept options can be generated and discussed with the operators on the basis of performance and cost. Typical results of parametric variations are shown in Figures 11.6 to 11.8. The more promising option(s) need now to be

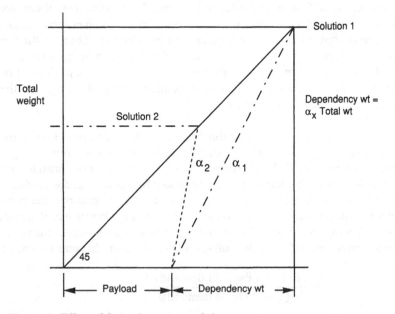

Fig. 11.5 Effect of dependency on total size

taken to a stage further in design detail to validate the concept and give assurance that the design of such a vessel could be carried forward to the next stage with little risk of requiring major revision. That stage we have termed as Feasibility Studies.

Given the sizing, configuration, performance and broad budgets established for the concept a more thorough investigation is now required. The main steps in the investigation are as follows:

(a) Structural calculations for pressure hull plating and framing, plus analysis of major junction details or possibly difficult structure geometries that have been identified.

(b) A more thorough analysis of resistance of the hull at the various operating speeds, e.g. maximum, snorting, submerged endurance speed.

(c) Analysis of the hydrodynamic propulsion efficiencies of selected propulsor, and hull interactions.

(d) More detailed analysis of propulsion machinery and battery performance.

(e) Outline diagrams of major system runs and location of major items of auxiliary machinery, e.g. HP air bottles, pumps.

(f) General arrangement drawings and compartment layouts. Weapon stowage, fore end and after end geometry.

(g) Estimates of size and location of control surfaces.

(h) Assessment and location of T & C Tanks plus location and capacity of other tanks.

(i) Based on the foregoing, re-assessment of weights and buoyancy elements, disposition of solid ballast, plus longitudinal balance submerged and floating trim on the surface.

(j) Assessment of hydrostatic stability submerged, surfaced and transition (possible surfacing in ice stability).

(k) Assessment of Weight Groups and formulation of Design Budgets.

(l) Assessment of Unit Cost and Through Life Cost.

THE FUTURE

11.40 The step by step process of design with increasing detail, through the stages of Concept Studies, Feasibility Studies, Design for Build, has evolved in an essentially manual environment of design. Computers have been introduced to speed up the calculation and drawing processes that each of the stages requires, but these have been primarily of the nature of a substitution of previous hand methods.

It is conceivable that with the increasing power of computers to handle large sets of data and manipulate graphics, the stage by stage process could be replaced by a single process. If suitable algorithms could be incorporated into the program a designer might be able to input the requirements and his choices of configuration, technology and standards, and for the computer to do the rest to produce a complete set of design data, drawings and specifications for building the submarine. That possibility may be a long way off but it is certainly conceivable that the more detailed design analysis of the Feasibility Stage might be incorporated so that, from the requirements and design selection input a Concept incorporating first stage Feasibility Assessment could be generated. That would amount to a telescoping of Concept and Feasibility into a single step in the design process. However, although one can postulate what might be possible in that regard, the question must be asked: is it desirable?

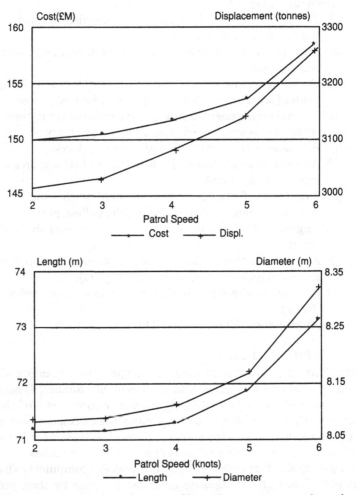

Fig. 11.6 Results of parametric survey: Size and costs versus patrol speed

This is not easy to answer. There is a real danger that, if an initial idea were taken too far down the design process, it could become too fixed. The designer and operational staff could be presented with a large amount of detail which might obscure the main issues. There is some sense in aiming to keep a broad canvas at the early stages whilst trying to decide which way to go. On the other hand the concept stage is for the consideration of many possible options and some assurance is required that each of them is about right. It is not necessary for it to be absolutely right but it must be relatively correct compared to alternative options. Thus if, at the concept stage, option A appeared better than option B this must remain so through the later stages of design even if both change somewhat when developed in more detail. For example, if A performed

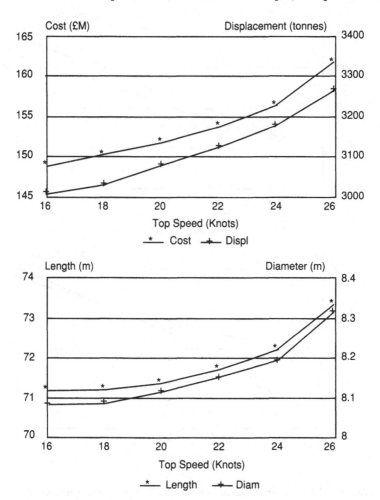

Fig. 11.7   Results of parametric survey: Size and costs versus top speed

all tasks and cost less than B which also satisfied the requirements, a decision to proceed with A would be likely, but that decision would be rendered invalid if subsequently the cost of A rose relative to B.

This discussion leads to a conclusion that perhaps the main analysis programmes of Feasibility should be incorporated into the Concept stage. Whilst the simpler algorithms of Concept would remain a means of obtaining a first estimate of size, the program would proceed into a cycle of more precise calculations normally left to Feasibility. For example, the simple estimate of pressure hull weight could go round a loop of more detailed design analysis and weight calculation. Similarly, the ini-

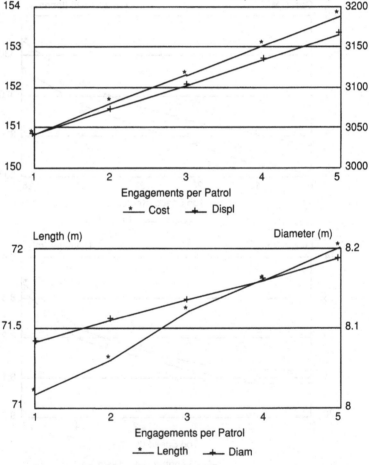

Fig. 11.8 Results of parametric survey: Size and costs versus number of engagements

tial estimate of power could be followed up with a more thorough analysis of resistance and propulsion efficiency.

In that way decisions on options could be made with much greater assurance than is presently possible, and the quantity of data output to be assessed would be much the same as at present.

In this book we have placed emphasis on the importance of achieving conceptual understanding of the processes at work and the inter-relations of the sub-systems within the design as a whole. There is a danger that, if the design process were mechanised too much, the designer would lose 'feel' for and understanding of the design and thereby his control of the process. The risk is particularly real for designers coming new to the activity.

There is still virtue in learning the hard way and undertaking at least one study by a manual method for those attempting to design for the first time.

APPENDIX 1 HYDROSTATIC CONDITIONS OF FLOTATION

For a vessel floating on the surface, the hydrostatic properties of relevance are the conditions of flotation and the stability of the vessel in relation to disturbances from the static flotation condition.

It is assumed that the water surface is calm and any movement of the craft is sufficiently slow for any dynamic effects to be discounted.

Accepting Archimedes principle that a vessel will displace its own weight of water, the initial requirement for flotation is that the intact hull is of sufficient volume to displace its own weight whilst having a reasonable freeboard (height of weather deck above the waterline) and that it floats upright. The volume of watertight hull above the waterline constitutes a Reserve of Buoyancy (ROB). This ROB becomes important when considering the safety of the vessel in damaged conditions when water floods part of the hitherto intact displacement volume below the waterline.

The stability of the vessel concerns the outcome of perturbations from the static flotation condition. Sideways motion (sway), change of heading (yaw) and fore or aft motion (surge) do not change the static conditions of the hull and can be considered as neutral. Whereas roll motion (heel), pitching and vertical motion (heave) result in changes in the distribution of buoyant volume and hence variation in the static equilibrium condition. The question to be answered is whether after a disturbance the vessel returns to its initial equilibrium state.

If a vessel heaves upwards then the displacement volume reduces and the excess of unchanged weight over reduced buoyancy provides a force in a direction restoring the vessel to its original position. The opposite occurs if the vessel is disturbed downwards. This is clearly, in a static sense, a stable motion. For a small heave motion the rate of change of buoyancy is primarily dictated by the area of the hull at the intersection with the water surface (WPA). The greater this area the larger the restoring force for a unit of heave motion. This simple result is somewhat modified by the vertical shaping of the hull at the waterline which can result in the WPA increasing or decreasing with heave departures from static equilibrium.

A pure forward pitching disturbance will cause the bow to go deeper and the stern to rise. This results in a change in the longitudinal distribution of buoyancy forces with more forward and less aft, which provides a restoring moment to the ship so that it will return to the initial conditions. The opposite situation arises for a stern pitch so this condition can also be considered as statically stable. The magnitude of the restoring moment will depend primarily on the longitudinal distribution of waterplane area. A ship with fine ends will have less relative pitch restoring moment than a vessel with full ends. The vertical changes in displacement volume also modify the response so that a vessel with flared bows will experience a strong restoring moment to large bow down pitch. Though they have been treated as separate motions the changes in waterplane due to heave and pitch interact so that strictly the stability must be considered as coupled motions.

The static stability of the ship in roll (heel) motion is somewhat more complex. In the equilibrium state the weight of the ship acts vertically through the centre of gravity whilst the equal opposing buoyancy force acts through the centre of the buoyant volume of displacement. At first sight the condition for static heel stability requires that the centre of gravity should be below the centre of buoyancy so that when the ship heels a positive restoring moment is applied to return the ship upright. This condition applies to a fully submerged submarine but is not a necessary requirement for a surface ship. As with the pitch motion, when a ship heels part of the hull on one side enters the water whilst on the other side part of the hull emerges from the water. The result is a lateral shift in the line of action of buoyancy force which provides a restoring moment acting against the heel. (Figure A1.1(a)) For small angles of heel it is found that the line of action of buoyancy forces acts through a point above the upright centre of displacement volume. This point is termed the transverse metacentre (M). Upright stability for small angles of heel therefore requires that the centre of gravity must be below the metacentre and not necessarily below the centre of displacement volume. This is usually defined as GM + ve, where GM is the vertical distance between the centre of gravity and the metacentre.

Whilst strictly speaking static stability should only apply to small perturbations about the static equilibrium of flotation, in Naval Architecture it is the practice to extend the considerations of heel stability to very large angles.

For large heel angles the concept of a metacentre for the line of action of buoyancy ceases to be useful. At some angle (30–40°) the edge of the weather deck goes below the waterline. The increase in restoring moment then starts to reduce. It reaches a maximum and then begins to fall and may become zero at a large angle. This angle is known as the angle of vanishing stability. A roll disturbance that causes the ship to pass this angle will result in capsize with the vessel, if still intact, finding a new equilibrium state upside down.

To present the roll stability characteristics of a vessel to large angles a different measure is used, which is represented as $\overline{GZ}$, which is defined as the normal distance between the parallel forces of weight and buoyancy at any angle. (Figure A1.1(b)) Clearly, when upright, the $\overline{GZ}$ value is zero and for small angles the relationship is that $\overline{GZ} = \overline{GM} \sin\phi$. At large angles, however, that relationship no longer holds.

The characteristics of the $\overline{GZ}$ curve depend on the shape characteristics of the hull. Large beam serves to increase the lateral shift of buoyancy and give a high initial metacentre. High freeboard increases the angle of deck edge immersion, delaying the turn over of the $\overline{GZ}$ curve, though there may be a rapid fall away to a relatively low angle of vanishing stability. A cross-sectional shape of particular interest to submarines is the circle. It can be shown that whatever the level or angle of the flotation waterline on a circular cross-section the metacentre always acts at the centre of the circle. (Figure A1.1(c)) Thus the modern submarine of circular section has a heel stability requirement that the centre of gravity should be below the circle centre when upright. Submerged the waterplane effect disappears but the centre of buoyancy is also near the centre of the circle so that $\overline{BG}$ is also positive. Because these points are invariant with angle, the $\overline{GZ}$ curve is a sine curve, $\overline{BG} \sin\phi$, which is positive to 180°. As discussed in Chapter 3, both the centre of gravity and the metacentre may migrate during the transition from submerged to surface and the above conditions may not then apply.

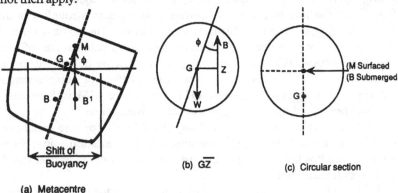

(a) Metacentre   (b) $\overline{GZ}$ (c) Circular section

Fig. A1.1 Measures of transverse stability

APPENDIX 2 OPERATIONAL PRACTICE FOR KEEPING IN TRIM

A2.1 CATCHING A TRIM

The process of accurately adjusting the state of a submarine to neutral buoyancy and longitudinal balance is termed 'catching a trim'. This is an important operation which is carried out when a submarine first dives when going on patrol, having left harbour and started passage on the surface. It provides a datum state for the submarine from which subsequent changes in weight or buoyancy can be easily compensated for, and enabling subsequent checks on trim made regularly during the patrol to be soundly based. The importance of keeping in trim is heightened should a submarine be required to remain quiet for any length of time because the process of catching a trim involves running pumps and so tends to be noisy.

It is desirable that when the submarine first dives at the start of a patrol it should not be grossly out of trim, and this requires that a good approximation to being in trim should be achieved before diving. The first step towards that end should have been taken when the submarine returned to harbour from its previous patrol, because it ought then to have been in good trim. From then on all changes in weight, e.g. due to re-storing, are carefully recorded and adjustments made to the contents of the Trim and Compensation (T & C) Tanks in order to correct for them. A methodical way often used for the purpose is by means of a 'trim crib', which has been previously drawn up for the submarine and gives the change in amount of water in the T & C Tanks appropriate to each weight change expected to be made during storing.

The state of trim of a submarine in harbour can be checked by observing the draughts at which it is floating and comparing them with what are known as the 'trim draughts', namely the draughts determining the condition of the surfaced submarine which would ensure that if it were to dive in the water in which it is floating it would be properly in trim. (The trim draughts are independent of the density of the water in which the submarine is floating on the surface). Any departure of the actual draughts from the trim draughts would indicate an out of trim condition, which could then be adjusted by appropriately changing the contents of the trim and compensation tanks until trim draughts were achieved.

A2.2 INITIAL DIVE

Assuming that the submarine has been brought reasonably close to a state of trim on first diving, it is maintained at a steady speed of say five to six knots in order to provide sufficient hydrodynamic forces to counter any residual out of trim which might still remain. Observation of the pitch angle on the submarine and the positions of the hydroplanes will indicate whether the boat is properly in trim. For a submarine which is correctly in neutral buoyancy and with longitudinal centres of gravity and buoyancy coinciding, the hydroplanes and boat attitude will be in what are known as their hydrodynamic balance positions, i.e. those at which the submarine can maintain a straight and level path independent of speed. These positions for the control surfaces are not necessarily their datum zeroes or even their zero lift settings, for, as we explain in the chapter on Submarine Hydrodynamics, the hull itself – particularly with bridge fin – experiences a bow-up hydrodynamic moment. To correct for this, the hydroplanes have to be set at balance angles to provide an opposing bow-down moment, which usually requires dive settings on both forward and after hydroplanes. In some submarines quite large balance angles, 4° or 5°, are called for and, as this requirement limits the hydroplane movement available for control action, it is sometimes the practice to assist hydrodynamic balance by proceeding with a small bow-down angle on the hull. As will be described later in the appendix an autopilot will automatically make this adjustment.

Once the submarine is under control dived, and at constant depth and level keel, speed is slowly taken off the boat. Variable ballast is transferred using the T & C System to maintain steady conditions if reducing speed shows up errors in either total weight/buoyancy relationship or longitudinal balance. It is customary to accept that a good trim has been caught if, at a speed of one or two knots, depth and angle control continue to be satisfactory. A stopped trim is more difficult to achieve because of the depth instability previously described. It should be recognised that the process of catching a trim would be made much more difficult if any air were trapped in tanks or free flood spaces, as this would greatly increase buoyancy changes with depth. Moreover, there is a risk that any air hold up in the tanks or casing, which did not escape during the slow motion movement involved in catching a trim, might be swept away when the boat increased speed, only subsequently to reveal that it was no longer in trim.

A2.3 SUBSEQUENT OPERATION SUBMERGED

The method described for catching a trim is standard practice for most submarines when first diving at the start of a patrol. For diesel electric submarines it is specially important to keep in trim as they spend a lot of their time submerged on battery power at relatively low speeds.

Submarines with nuclear propulsion plant do not, however, spend much time at slow speed submerged, and so there could be quire a long interval of many days before another opportunity arose to catch a trim.

Nuclear submarines, particularly since they have steam plant, tend to experience quite large changes in weight on patrol because of the relatively large crew (by submarine standards) and the associated consumption of stores and also because of the requirement to make distilled water, both for the crew and the steam boilers. The latter feature causes the brine tanks to fill with strong saline solution which the plant engineers regularly dump to sea, so altering trim. At speed, the effect on trim would not be very pronounced as the hydrodynamic forces available would largely swamp the hydrostatic changes. In consequence, a nuclear submarine proceeding at quite high speeds for protracted periods could become substantially out of trim. Since it might not be desired by the command to slow down in order to correct the trim, an alternative way of keeping the state of trim is required so that corrective action can be taken. Observation of the behaviour of the autopilot serves that purpose. We dealt in Chapter 8 with the logic of the autopilot, and for now just comment that it is possible to use the depth and pitch angle solutions which the autopilot is capable of providing to judge the state of trim of the submarine. Combinations of pitch angle on the boat and the angles on the forward and after hydroplanes at moderate speeds serve to indicate with reasonable accuracy how light or heavy the submarine has become and with what amount of out of balance in longitudinal moment. (The method is applicable at all speeds, but at higher speeds the angle might be so small as to be difficult to observe). It is because of these angles on the boat and the hydroplanes, to support out of balance hydrostatic force and moment by hydrodynamic forces and moments, that the drag is increased.

By way of illustration, if the submarine were heavy without much in the way of out of balance moment, the autopilot would find a solution in which the hull would be at a bow-up angle to support the extra weight by hydrodynamic force and the hydroplanes at dive angles so as to counter the bow-up hydrodynamic moment due to the hull. By the same reasoning, if the submarine were light with little out of balance moment, the solution would put the hull at a bow down angle and the hydroplanes to rise angles. If in either case there were also an out of balance longitudinal moment, the autopilot would find other solutions in which the hydroplane angles would be appreciably different. But in all cases it would be apparent to the officer of the watch not only that the submarine was out of trim but also approximately by how much and in what way. This alternative is not precise, but in conjunction with a computer programmed for the purpose can provide continuous indication of state of trim and what corrections by the T & C system would be necessary to restore trim.

APPENDIX 3 ASSESSING WEIGHT AND SIZE OF VARIABLE BALLAST

In design, when the trim and compensation arrangements are being addressed, a procedure is required for sizing the tanks and assessing the amount of water in them, i.e. the weight of variable ballast. The method we now describe, which is widely used for the purpose, is to be read in conjunction with Figure A3.1.

(a) A standard condition of the submarine is defined, i.e. one with full crew, stores and fuel on board, with the boat in water of standard density 1.0275. A statement is drawn up of all the weights and locations of items which might vary during the course of a patrol, including buoyancy changes, especially those associated with extremes, e.g. maximum changes of sea-water density between say 1.00 and 1.03 (or whatever is appropriate for the submarine under design).

(b) A graph is prepared with a horizontal axis for longitudinal moments and a vertical axis for forces, the origin representing the standard condition of the boat. The weight and buoyancy changes in the statement from (a) are plotted as a series of discrete points on the graph to the convention that reductions in weight forward are placed in the first quadrant. The aim is to cater for demanding combinations of weight and buoyancy changes which are sensible but nevertheless still serve to establish the boundary of extreme conditions in which the boat might find itself. Thus towards the end of a patrol a submarine, having consumed nearly all of its stores and fuel, might have to enter an area in which sea-water density is high; alternatively a submarine shortly after leaving base to go on patrol might have to enter water which is almost fresh. It is usual to find that most of the extreme combinations lie either in the first quadrant (i.e. light forward due to the consumption of stores) or in the third quadrant (i.e. heavy slightly aft due to the buoyancy change with little weight reduction).

(c) The extreme points of this plot indicate the range that has to be catered for by T & C tanks. To establish how this can be done, a second graph to the same scale as the first is prepared to represent the capacities of the proposed tanks. The procedure is to start from the origin and plot into the first quadrant the effect of flooding the foremost empty trim tank, which will increase weight upwards on the plot and the forward moment (about amidships) to the right on the plot; subsequent tanks working from forward to aft are progressively flooded until finally the after trim tank is full. The boundary thus far obtained defines the maximum weight change which can be catered for by the proposed tanks.

(d) Having reached this maximum condition on the second graph (which has probably moved the moment slightly aft) the process is put into reverse by emptying the forward trim tank and then working back through the boat until all tanks have been emptied in sequence. This procedure provides a polygonal 'lozenge' which represents the envelope of the maximum variations in weight and moments achievable with the proposed tanks.

(e) The tank polygon of the second graph is now overlaid on the weight and buoyancy changes of the first plot and its position adjusted in an attempt to encompass all the extreme points identified in the first graph. If no satisfactory position can be found, it is necessary to modify the polygon by changing the capacity of the tanks until a solution is obtained. The final polygon then defines the required capacities of the T & C tanks. Moreover, the location of the origin of the first plot in relation to the second identifies the standard trim condition of the submarine, i.e. the amount of weight and moment of the water that has to be disposed in the tanks, which enables them to be emptied if an extremely heavy condition were encountered on patrol while ensuring that there is enough space in the tanks to enable them to be filled if an extremely light condition were encountered. The standard condition of T & C, as it is termed, is the contribution of that system to the amount of variable ballast to be included in the weight estimate.

At the preliminary stages of design the details required for the foregoing will not be available. Nevertheless, an allowance for the volume occupied by T & C tanks must be made to obtain a required volume of the pressure hull. It will be noted that the extreme weight variations arise from:

(a) fully stored in low density water (extreme heavy)

to

(b) end of patrol stores in high density water (extreme light).

Thus an estimate of T & C tank volume can be made by the following dependency algorithm:

Vol of tanks =

$$\left[\text{Ph Vol} \times \text{range of density \%} + \frac{\text{Weight of variable stores}}{\rho_w} \right]$$

$\times$ Utility factor

The utility factor is of order 1.1 to allow for internal tank structure and residual water that cannot be pumped or blown due to geometry of the tanks.

Figure A3.2 shows a typical Trim Polygon produced in the course of a Submarine Design Exercise at University College, London.

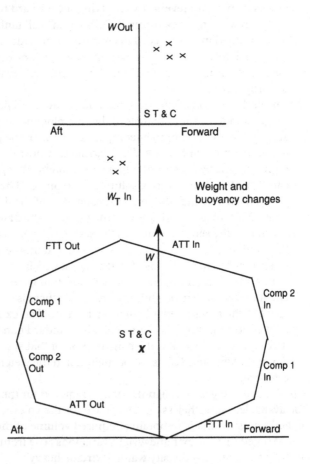

Fig. A3.1 Scope of trim and compensation tankage

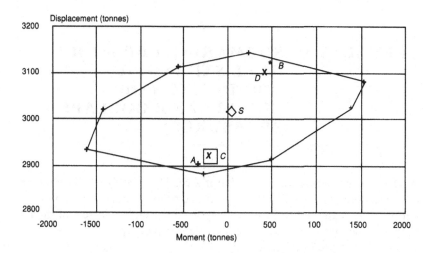

Fig. A3.2 Trim Polygon

APPENDIX 4 SUBMARINE PRESSURE HULL STRENGTH (PREDICTION OF INTERFRAME COLLAPSE PRESSURES)

A4.1 It can be readily shown that if a vessel in the form of a circular cylinder, domed at its ends, of radius R and shell thickness t has an internal pressure p applied to it, then away from the ends two sets of membrane stresses will be set up in the shell plating, namely:

$\sigma_1 = pR/2t$ longitudinally

and $\sigma_c = pR/t$ circumferentially.

Because the circumferential stress is double the longitudinal stress, as the internal pressure is progressively increased the cylinder will fail by starting to yield circumferentially when the hoop stress σ_c approaches the yield strength of the shell material, σ_y It is usual to describe the pressure to initiate yield as the 'boiler' pressure: $p_b = \sigma_y \cdot {t/R}$.

A4.2 Turning now to what would happen if the pressure were applied externally to a long cylinder, if it was unsupported by ring stiffeners, it would fail well away from the ends by buckling elastically in an oval mode at a pressure several orders of magnitude lower than the boiler pressure – even if it were perfectly circular to start with. This behaviour is amenable to analysis, and a formula for the elastic collapse pressure of an unstiffened circular cylinder with bulkheads spaced a distance L apart was derived by a German researcher, von Mises, in the early 1920s. His work showed that if L was commensurate with R, i.e. the cylinder was relatively short, the elastic collapse mode would not be oval but of a higher order with multiple lobes; even then the von Mises collapse pressure, which we here designate as p_m, would be very much lower than p_b.

A4.3 It transpires that the ratio p_m/p_b provides a useful measure for a short unstiffened cylinder of its proneness to failure by elastic buckling in relation to its potential yield strength. The ratio is particularly helpful to understanding what can be achieved by introducing ring frames to support the cylinder against external pressure. Although a ring frame is less rigid radially than a bulkhead, the von Mises formula can be applied to a good approximation to indicate the pressure to cause elastic buckling of the shell plating between adjacent ring frames. For typical submarine pressure hull geometries, the cylinder over the length between adjacent frames is effectively very short and so the elastic buckling mode is one with many lobes, and the corresponding values of p_m relatively high. It is possible to raise the interframe collapse pressure to within about 10% of p_b by a relatively modest investment of weight in ring frame stiffening of the pressure hull. The formula for p_m involves the modulus of elasticity for the shell plating material, the radius and thickness of the shell plating and the ring frame spacing – but not the frame size – and it is by manipulation of just those parameters that p_m can be influenced.

A4.4   Before going on to explain how the favourable outcome of increased interframe collapse pressure can be achieved, we take a step backwards in time to the early days of submarine pressure hull design, long before there was any prospect of evaluating interframe collapse pressures by computer-based numerical methods. Submarine structural researchers needed to have means to take account of the effects of fabrication distortions and residual stresses on the results of the theoretical methods of analysis then available, which necessarily excluded such effects. The natural recourse was to test structural models of the geometry of interest under hydrostatic pressure, i.e. to proceed empirically. Because the concern was with interframe collapse, the models were relatively short, rarely more than a few frame bays long, and had reinforced ends to prevent failure there – where the longitudinal stresses were unnaturally high. Over the years, the results of independent tests on many models from many countries and of widely different geometries have become openly available – the total currently is well over 700 – and so there has been a search to find a non-dimensional way of presenting the results in as useful a form for designers as can be contrived.

A4.5   A somewhat primitive form of non-dimensional presentation of the empirical interframe collapse data, which the authors were associated with developing several decades ago, is illustrated in Figure A4.1. For each model tested, the pressure, p_c, to cause collapse was noted, the values of p_b and p_m were calculated from the model's known geometry and yield strength, and the result of that particular test was plotted on the diagram having evaluated the ratios p_c/p_b and p_m/p_b. Also shown on the diagram in Figure A4.1 are two lines, a slant line through the origin to represent $p_c/p_b = p_m/p_b$ and a horizontal line at a height $p_c/p_b = 1.0$. Those two lines can be regarded as the asymptotes to which the collapse pressure results for models of extreme geometry will tend. Along the diagonal line near to the origin, $p_c = p_m$ because in that regime the ring frames are too widely spaced to avoid failure by elastic buckling of the shell plating between the frames; a pressure hull of that geometry would be very inefficient structurally, and that is reflected by the low values of p_m/p_b

involved. Along the horizontal line well away from the origin $p_c = p_b$ because the ring frames are sufficiently close together to avoid failure by elastic buckling of the shell plating between frames; failure is due to the hoop stress in the shell plating there approaching the yield strength of the material, which is quite efficient structurally.

A4.6   Several significant features can be identified from the empirical data presented in that way. Pressure hull models with relatively low values of p_m/p_b have collapse pressures which show considerable scatter because they are particularly prone to geometric inaccuracy in manufacture. On the other hand convergence to the horizontal asymptote at relatively high values of p_m/p_b is good because models of that sort of geometry are not at all sensitive geometrically. Models in between those extremes effect the transition from diagonal to horizontal with less scatter as the liability to elastic buckling failure is reduced. It can be appreciated that a law of diminishing returns applies to the results in the sense that – bearing in mind the increase in pressure hull weight which accompanies increase in p_m/p_b value – there is progressively less benefit in approaching the horizontal asymptote.

A4.7   Improvement in this form of presentation has concentrated on finding a better measure than the boiler formula value for the hoop stresses in a ring frame stiffened cylinder. As observed in Section 5.13, the use of ring frame stiffeners to support the shell plating of a pressure hull against premature buckling causes a complex pattern of stress in the plating due to the variation in circumferential strain from that in way of the frames to that in between the frames. In consequence, the largest hoop stress occurs midway between frames which – because of the support from the frames – is actually lower than the boiler formula value, $\sigma_c = pR/t$. An adequate theoretical treatment of the stress distribution in the shell plating of ring frame stiffened cylinders was, in fact, derived as long ago as the late 1920s by two German researchers, von Sanden and Gunther, which is algebraically complicated but amenable to numerical analysis. Preference for an alternative to p_b in the application to the non-dimensional presentation of the empirical data mentioned earlier has settled on p_y, which designates the external pressure to cause the mean hoop stress at the mid-thickness of the shell plating midway between frames based on von Sanden and Gunther to equal the yield strength of the shell material. This alternative approach to presenting the empirical data for collapse of pressure hull models under external pressure, which is now widely used, plots the ratio p_c/p_y to base p_m/p_y. That is the form of the diagram in Figure 5.7, and it can be seen that the significant features described above for the diagram using p_b are retained. The improvement lies in the greater realism that follows from the use of a closer representation of the hoop stress pattern in the shell plating of a ring frame stiffened cylinder.

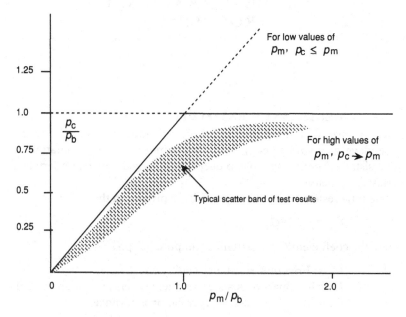

Fig. A4.1 Empirical plot of model pressure test results

APPENDIX 5 ESTIMATES OF RESISTANCE AND PROPULSION

RESISTANCE ESTIMATES

Once the size of the submarine has been established and the initial appendage geometries determined then a more detailed evaluation of the resistance/speed relationship deeply submerged may be undertaken as described below.

The total resistance coefficient may be expressed in the form:

$$C_{TS} = C_{FS} + C_R + C_A$$

where the coefficient C = Resistance component/$\frac{1}{2}\rho SV^2$

and C_{FS} = Frictional resistance coefficient
        C_R = Residuary resistance coefficient (mainly viscous pressure)
        C_A = Model to full-scale correlation allowance

The value of C_R is determined from model resistance experiments and is assumed to be independent of Reynolds number i.e. $C_{RM} = C_{RS}$. Errors implicit in this assumption are accommodated in the correlation allowance.

The correlation allowance is determined ultimately from First of Class trials for a new submarine. However, at the design and model experiment stage a suitable value of C_A is assumed, based on data from previous classes of submarine. A typical value of the correlation allowance might lie between 0.4×10^{-3} and 0.6×10^{-3}.

There is considerable variation in both the residuary resistance coefficient determined from model experiments and in the correlation allowance deduced from trials data. Fortunately, however, the data indicates that their sum is less variable than either of the individual components. This may be partly explained by the fact that forms with a high L/D ratio have a lower C_R (i.e. less form drag), but possess a larger wetted surface area for a given volume of displacement. The correlation allowance takes account of the effects of hull roughness (not fouling), hull openings, building tolerances, and errors inherent in the scaling procedure. Consequently forms with large wetted surface areas for a given volume of displacement will have larger correlation allowances. In adding C_R to C_A therefore these effects tend to cancel.

This may be seen from Figure A5.1, which displays a small range of (C_R+C_A) values $(0.9 - 1.2) \times 10^{-3}$ for a large range of slenderness ratios and block coefficients. Figure A5.1 can be used for the final design evaluation.

The frictional resistance coefficient C_{FS} for the submarine is calculated using an empirical formula (the 1957 ITTC Line):

$$C_{FS} = \frac{0.075}{(Log_{10}R_N - 2)^2}$$

The hull and casing and each appendage are treated separately using the local Reynolds numbers based on the overall length for the hull plus casing and the mean chord length of each appendage. The individual coefficients are then added, each contribution being in the ratio of its area to the total wetted surface area (known as Reference Area, S_{REF}), i.e.:

$$C_{FS} = C_{FHull} \times$$

$$\frac{\left(\text{Wetted area of hull} - \text{area covered by appendages} \right)}{S_{Ref}}$$

$$+ C_{FBridge1} \times \frac{\text{Wetted area of bridge}}{S_{Ref}}$$

$$+ 2\, C_{FBowplane} \times \frac{\text{Wetted area of one bowplane}}{S_{Ref}}$$

$$+...etc.$$

where S_{Ref} = Wetted area of bare hull and casing plus wetted area of all appendages minus the area covered by appendages.

The total ship resistance coefficient having been determined, the effective power (P_E) needed to 'tow' the submarine at speed V in the submerged condition is given by:

$$P_E = \tfrac{1}{2}\,\rho.V^3.S_{Ref}.C_{TS}$$

An allowance may be made for hull fouling in temperate waters by the addition of an extra $\tfrac{1}{8}\%$ C_{FS} per day out of dock up to a limit of 180 days.

A knowledge of the propulsive coefficient PC would now enable the shaft power P_S to be determined across the speed range, but before discussing that step we comment on how the resistance of the submarine is affected by the depth at which it operates.

Near-surface and surface resistance

As the submarine approaches the surface its resistance increases due to the dissipation of energy by the waves then generated. The creation of the waves modifies the pressure distribution over the submarine so that the form drag is increased and this manifests itself as an increase in the residuary resistance coefficient, δC_R.

The increase is greatest as the hull starts to break surface, that is, when the draught (H) equals the depth of the hull $(D = $ hull diameter plus casing height). For this condition $(H = D)$ the increase in residuary resistance coefficient may be related to Froude number (F_n) and a non-dimensional parameter $(L/\nabla_{Form}^{1/3})$. Typical data are shown in Figure A5.2.

As the submarine surfaces further from this condition a *very crude estimate* of surface resistance may be made by assuming that C_F, C_A, C_R and δC_R are constant and the reduction in resistance which occurs is due solely to the change in wetted surface area.

As the submarine goes deeper δC_R is reduced until the effect of the free surface becomes negligible. Although only limited data is available on the variation of wavemaking resistance with depth and speed, the following formula may be used to give an approximate value for the increase in residuary resistance coefficient:

$$\delta C_R^{''} = \delta C_R \; e^{\left[0.37g \, \frac{(D-H)}{V^2} \right]} \text{ for } H > D.$$

The resistance due to periscopes and other masts has to be assessed separately. The close proximity of a number of in-line and/or sided masts makes prediction very difficult. A resistance coefficient of 0.3 is recommended, based on the frontal area of each mast.

PROPULSION ESTIMATES

(a) Initial propulsive coefficient estimate (submerged)

The interaction of the propeller and the hull necessitate the design of the aft end of the submarine and the propeller being carried out together. Investigations into the effects of after-body fullness and propeller diameter have shown how the hull efficiency elements and the resulting propulsive coefficient vary in consequence. Increasing the fullness of the after-body results in increased resistance and wake. The increased wake can be used to increase the hull efficiency such that shaft power is not significantly affected. However, full after-bodies are a definite disadvantage in terms of rudder and sternplane effectiveness especially if there are span constraints.

An increase in propeller diameter offers a potential improvement in propulsive efficiency. However, as the ratio of the propeller to hull diameter is increased the average wake over the propulsor is reduced, especially if the tips of the propeller are working outside the wake field as they will be with large diameters.

The submerged propulsive coefficient for single screw submarines may vary between say 0.7 and 0.8, depending upon the relative size of the propeller, the shape of the after-body, the rpm characteristics of the machinery and the propeller design. At the early concept design stage a value of 0.75 for single screw submarines, and 0.65 for multi-screw submarines would not be unreasonable for use in the estimation of shaft power, (P_S).

(b) More detailed propulsion estimates (submerged)
(i) Propeller efficiency

Open water characteristics for a propeller design may be determined from the AEW 20 inch Methodical Series. Current submarine propellers differ considerably in the number and shape of the blades from the propellers used in determining the Series Data, the developments being aimed at improving the noise performance while maintaining thrust. However, in the non-cavitating region the propulsive characteristics of modern propellers may be assumed to be similar to the Series Data. Propeller blade loading should be limited to 70 kN/m².

(ii) Hull efficiency elements

The hull efficiency elements are primarily dependent upon the shape of the after-body and the size of the propeller relative to the hull. Typical curves of Taylor wake (w_T), Thrust deduction factor (t) and Relative rotative efficiency (η_R) are presented in Figures 6.7, 6.8 and A5.3 respectively.

Values of w_T presented in Figure 6.7 are typical model values. Full-scale wake values are somewhat lower than the model values as a result of scaling difficulties (boundary layers do not scale). A reduction in the model wake values of the order of 12% appears to be a reasonable correction.

(iii) Propulsive coefficient

'Ideal' trials factors should be used so that the Quasi-propulsive coefficient factor can be taken as being equal to the shafting efficiency (η_S), which is usually taken as 0.98, and T_R/T_M can be taken as equal to 1.0.

Consequently the propulsive coefficient may be evaluated from the formulae:

$$PC = \frac{P_E}{P_S} = \eta_H\,\eta_O\,\eta_R\,\eta_S$$

(c) Near or on surface propulsion estimates

Little data is available on propeller characteristics and hull efficiency elements for submarines at periscope depth or on the surface. It is quite reasonable to assume that the PC reduces from that estimated for the submerged condition. Reductions of the order of 7 and 14% respectively would not be too unrealistic for single screw submarines.

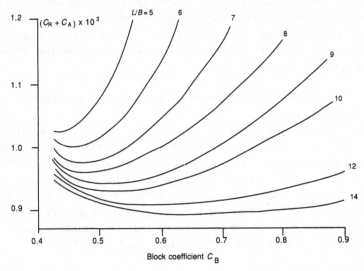

Fig. A5.1

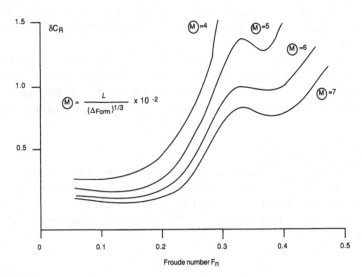

Fig. A5.2

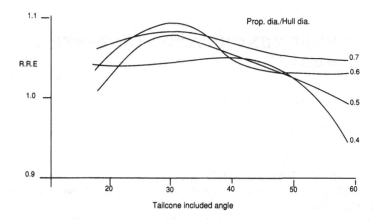

Fig. A5.3 Variation in relative rotative efficiency

REFERENCES AND SUGGESTED READING

GENERAL DESIGN
E.S. Arentzen, and P. Mandel, *Naval Architecture Aspects of Submarine Design.* SNAME (1960).

N. Friedman, *Submarine Design and Development.* Conway Maritime Press, (1984).

U. Gabler, *Submarine Design.* Bernard & Craefe Verlag, (1986).

N. Polmar, *The American Submarine.* The Nautical and Aviation Publishing Company of America, Anapolis, Maryland, 2nd Edition, (1983).

P.G. Wrobel, Design of the Type 2400 Patrol Class Submarine. *Trans. RINA,* (January 1985).

HISTORY
R. Baker and L.J. Rydill, *The Building of the Two Dreadnoughts.* Proceedings of the Third Shipbuilding Conference, National Maritime Museum, (1983).

R. Compton-Hall, *Submarine Boats – The Beginnings of Underwater Warfare.* Conway Maritime Press, (1983).

R. Compton-Hall, *Submarines and the War at sea 1914-18.* Macmillan, (1991)

R. Compton-Hall, *The Underwater War 1939-45.* Blandford Press, (1982).

R. Compton-Hall, *Submarine Warfare – Monsters and Midgets.* Blandford Press, (1985).

E. Rossler, *The U-boat, Technical Details.* Arms & Armour Press, (1981).

V.E. Tarrant, *The U-boat Offensive 1914-1945.* Arms & Armour Press, (1989).

J. Terraine, *Business in Great Waters.* Leo Cooper, (1989).

STRUCTURES
D. Faulkner, The safe design and construction of steel spheres and end closures of submarines habitats and other pressure vessels. *Proceedings BOSS,* (1979).

S.B. Kendrick, *The Buckling Under External Pressure of Ring Stiffened Circular Cylinders.* RINA, (1965).

S.B. Kendrick, Design of submarine structures. *Advances in Marine Structures,* (1986).

HYDRODYNAMICS
M.A. Abkowitz, *Stability and Motion Control of Ocean Vehicles.* MIT Press (1969).

E. Dempsey, Static stability characteristics of a systematic series of stern control surfaces on a body of revolution. *DTNSRDC Report 77-0085,* (August 1977).

J. Feldman, DTNSRDC revised standard submarine equations of motion. *DTNSRDL/SPD-0393-09*, (June 1979).

M. Gertler and G. Hagen, Standard equations of motion for submarine simulation. *NSRDC Report 2510*, (June 1967).

A.O. Hess, *On the Problem of shaping an Axisymmetric Body to Obtain Low Drag at Large Reynolds Numbers*. SNAME, (1976).

M. Gertler, Some recent advances in dynamic stability and control of submerged vehicles. *International Symposium, IMechE*, (1972).

J.B. Spenser, Stability and control of submarines. *JRNSS, Vol. 23, No. 3,* (1967).

AIP SYSTEMS

V.W. Adams, Fuel cells and possible naval applications. *JNS., Vol. 12, No. 2,* (1986).

Three International Symposia were organised by the Royal Institution of Naval Architects in 1983, 1988 and 1991. The following are some selections of papers presented.

1983 *International Symposium on Naval Submarines*, London.
R.J. Daniel, Considerations influencing submarine design.
D. Faulkner, The collapse strength and design of submarines.
H.A. Jackson, Submarine parametrics.
D. Wilson, Submarine escape and rescue, a review of recent developments in the Royal Navy.

1988 *International Symposium on Naval Submarines*, London.
G. Boisiayon, Design criteria for conventional submarines.
R.K. Burcher and C.V.Betts, The teaching of submarine design
B. Cederholm, CAD in submarine design
R.J. Daniel, Submarines Today
A.J. Giddings, Overcoming submarine control surface forces and flooding casulalties.
R.S. Gosling, *et. al.*, The AMPS nuclear reactor AIP source for diesel/electric submarines
M. MacKay, The prediction of submarine out of plane force and moment using Panel Method.
B Witterind and B.Wibbels, A closed cycle diesel propulsion system for submarines.

1991 *International Symposium on Naval Submarines*, London.
R. Dahlinder, A realistic air independent propulsion system for today and the near future.
G.H. Fuller, Submarines ready for their second century.
J. L. Grepel, A. G. Bovis, New trends in the design of steering and diving control systems.
L. Truedssen, Experiences of use of CAD systems in submarine design.

INDEX

Printed in the United States
By Bookmasters